GW00832079

Molecular Biology Intelligence Unit

Rb and Tumorigenesis

Maurizio Fanciulli, Ph.D.
Laboratory "B"
Regina Elena Cancer Institute
Rome, Italy

Landes Bioscience / Eurekah.com
Georgetown, Texas
U.S.A.

Springer Science+Business Media
New York, New York
U.S.A.

Rb and Tumorigenesis

Molecular Biology Intelligence Unit

Landes Bioscience / Eurekah.com
Springer Science+Business Media, Inc.

ISBN: 0-387-32173-X Printed on acid-free paper.

While the authors, editors and publisher believe that drug selection and dosage and the specifications and usage of equipment and devices, as set forth in this book, are in accord with current recommendations and practice at the time of publication, they make no warranty, expressed or implied, with respect to material described in this book. In view of the ongoing research, equipment development, changes in governmental regulations and the rapid accumulation of information relating to the biomedical sciences, the reader is urged to carefully review and evaluate the information provided herein.

Springer Science+Business Media, Inc., 233 Spring Street, New York, New York 10013, U.S.A.
http://www.springer.com

Please address all inquiries to the Publishers:
Landes Bioscience / Eurekah.com, 810 South Church Street, Georgetown, Texas 78626, U.S.A.
Phone: 512/ 863 7762; FAX: 512/ 863 0081
http://www.eurekah.com
http://www.landesbioscience.com

Printed in the United States of America.

9 8 7 6 5 4 3 2 1

Library of Congress Cataloging-in-Publication Data

Rb and tumorigenesis / [edited by] Maurizio Fanciulli.
p. ; cm. -- (Molecular biology intelligence unit)
Includes bibliographical references and index.
ISBN 0-387-32173-X (alk. paper)
1. Retinoblastoma. I. Fanciulli, Maurizio. II. Series: Molecular biology intelligence unit (Unnumbered).
[DNLM: 1. Genes, Retinoblastoma--physiology. 2. E2F Transcription Factors. 3. Neoplasms--etiology. 4. Retinoblastoma Protein--physiology. WW 270 R279 2006]
RC280.E9
616.99'484--dc22

2006002239

Dedication

In memory of Raffaele "Peo" Tecce

CONTENTS

EDITOR

Maurizio Fanciulli
Laboratory "B"
Regina Elena Cancer Institute
Rome, Italy

CONTRIBUTORS

Steven P. Angus
Department of Molecular Genetics and Microbiology
Howard Hughes Medical Institute
Duke University Medical Center
Durham, North Carolina, U.S.A.
Email: Steven.Angus@duke.edu
Chapter 3

Grazia Camarda
Department of Environment and Primary Prevention
Istituto Superiore di Sanitá
Rome, Italy
Chapter 2

Srikumar P. Chellappan
Drug Discovery Program
Department of Interdisciplinary Oncology
H. Lee Moffitt Cancer Center and Research Institute
Tampa, Florida, U.S.A.
Email: ChellaSP@moffitt.usf.edu
Chapter 5

Chantal E. Crémisi
Oncogenèse, Différenciation et Transduction du Signal
IRC-CNRS
Institut Fédératif André Lwoff
Villejuif, France
Email: cremisi@vjf.cnrs.fr
Chapter 1

Marco Crescenzi
Department of Environment and Primary Prevention
Istituto Superiore di Sanitá
Rome, Italy
Email: crescenz@iss.it
Chapter 2

Laetitia Daury
Laboratoire de Biologie Moléculaire Eucaryote
CNRS UMR 5099
Institut d'Exploration Fonctionnelle du Génome
Toulouse, France
Chapter 7

Fred Dick
London Regional Cancer Centre
Children's Health Research Institute
and the University of Western Ontario
London, Ontario, Canada
Chapter 6

Nicholas Dyson
Laboratory of Molecular Oncology
Massachusetts General Hospital Cancer Center
Charlestown, Massachusetts, U.S.A.
Email: dyson@helix.mgh.harvard.edu
Chapter 6

Peggy J. Farnham
McArdle Laboratory for Cancer Research
University of Wisconsin
Medical School
Madison, Wisconsin, U.S.A.
Email: farnham@oncology.wisc.edu
Chapter 4

Kerry L. Ferguson
Ottawa Health Research Institute
University of Ottawa
Ottawa, Ontario, Canada
Chapter 8

Erik S. Knudsen
Department of Cell Biology
College of Medicine
Vontz Center for Molecular Studies
University of Cincinnati
Cincinnati, Ohio, U.S.A.
Email: erik.knudsen@uc.edu
Chapter 3

Lucia Latella
Dulbecco Telethon Institute
c/o Parco Scientifico Biomedico
di Roma, San Raffaele
Rome, Italy
Email: llatella@dti.telethon.it
Chapter 9

Estelle Nicolas
Laboratoire de Biologie Moléculaire
Eucaryote
CNRS UMR 5099
Institut d'Exploration Fonctionnelle
du Génome
Toulouse, France
Chapter 7

Jaya Padmanabhan
Department of Biochemistry
University of South Florida
Tampa, Florida, U.S.A.
Chapter 5

Deborah Pajalunga
Department of Environment
and Primary Prevention
Istituto Superiore di Sanitá
Rome, Italy
Chapter 2

Linda L. Pritchard
Laboratoire Oncogenese
Differenciation et Transduction
du Signal
CNRS UPR 9079
Institut Andre Lwoff
Villejuif, France
Chapter 1

Pier Lorenzo Puri
Dulbecco Telethon Institute,
c/o Parco Scientifico
Biomedico di Roma, San Raffaele
Rome, Italy
Email: plpuri@dti.telethon.it
Chapter 9

Ruth S. Slack
Department of Cellular
and Molecular Medicine
Ottawa Health Research Institute
University of Ottawa
Ottawa, Ontario, Canada
Email: rslack@uottawa.ca
Chapter 8

Didier Trouche
Laboratoire de Biologie Moleculaire
Eucaryote
CNRS UMR 5099
Institut d'Exploration Fonctionnelle
du Génome
Toulouse, France
Email: trouche@ibcg.biotoul.fr
Chapter 7

Jacqueline L. Vanderluit
Ottawa Health Research Institute
University of Ottawa
Ottawa, Ontario, Canada
Chapter 8

PREFACE

The retinoblastoma gene product (pRb) and the closely related proteins p107 and p130 are central regulators of cell homeostasis, involved in the control of such critical functions as proliferation, differentiation and apoptosis. Indeed, the inactivation of Rb gene is implicated in a wide variety of human tumors, including familial retinoblastomas and osteosarcomas, as well as sporadic lung, prostate, bladder and breast carcinomas. Moreover, components of the Rb cell cycle regulatory pathway are altered in almost all cancers. On the basis of more than a decade of studies, it is now almost universally accepted that the activity of the transcription factor E2F is inhibited by its interaction with Rb, and the release of E2F by the viral oncoproteins would be a key event responsible for both their oncogenic properties and cell cycle progression and proliferation. However, a growing body of evidence suggests that the role of Rb in the cell may be more complex, indicating its involvement in the maintenance and induction of a differentiated phenotype, in cell survival and in senescence.

This book describes new insights into retinoblastoma gene family functions. Many of the chapters focus on the emerging new roles of pRb, exerted not only by regulating E2F family transcription factors. Peggy Farnham describes in her chapter how pRb cooperates with site-specific transcription factors to activate transcription, suggesting mechanisms by which Rb can function to positively regulate transcription. The chapter by Knudsen and Angus illustrates how pRb participates in the regulation of DNA replication, interacting with several important proteins involved in the control of this process.

Recent advances have demonstrated the interaction of pRb with chromatin remodeling enzymes and the potential roles of these interactions in pRb functions, providing some evidence that distinct pRb co-repressor may target different genes in different phases of the cell cycle. Therefore, much of the pathological gene silencing that occurs in cancer can be the consequence of the mistargeting of these enzymes on Rb. The chapter by Trouche et al extensively reviews these results and the involvement of histone modifying activities in the regulation of E2F-responsive promoters. In addition, the chapter by Dick and Dyson examines the current models of E2F activity in cell cycle control, but it also highlights many of the exciting new insights into E2F activity in development and apoptosis. Other chapters describe other kinases capable of functionally inactivating pRb in response to multiple stimuli (Chellappan), or highlight the role of the Rb family members in the developing embryo (Vanderluit et al). Three chapters focus on the role of pRb as a positive regulator of different differentiation processes (Pajalunga et al, Crémisi and Pritchard, and Latella and Puri).

I thank all of those colleagues who have brought us where we now stand, and I hope that this book will help point the way for those who will continue to advance in the ever more intriguing field of pRb cell biology regulation.

Maurizio Fanciulli, Ph.D.

CHAPTER 1

RB as a Positive Transcriptional Regulator during Epithelial Differentiation

Chantal E. Crémisi* and Linda L. Pritchard

Abstract

RB plays an essential role in epithelial cell differentiation and viability, these two properties being totally linked and independent of p53. To exert these functions, RB acts as a positive transcriptional coregulator, being recruited to the native gene promoters by sequence-specific transcription factors such as AP-2, thus implying direct activation of the target genes, rather than the downregulation of a repressor. Physical and functional interactions have been shown to exist in vivo between RB and transcription factors such as AP-1, AP-2 and SP1 family members, and a number of RB target genes that are specifically activated by RB in epithelial cells have been identified, including c-jun, collagenase, E-cadherin, p21 and Bcl-2. It is likely that other proteins are also associated with the RB/transcription factor protein complexes – in particular, proteins with histone acetyltransferase (HAT) activity, because gene promoters were found to be specifically acetylated when RB and AP-2 bound to them. Since comparable results have been reported for an osteoblast differentiation model, it seems likely that this mechanism might constitute a new paradigm for RB action in several differentiation systems. The mechanism of interaction of RB with viral oncoproteins seems to be different when it acts as a positive regulator versus a negative regulator. In differentated epithelial cells, the RB trancription factor complex is not dissociated by oncoprotein such as SV40LT, but rather a tripartite complex is formed containing the oncoprotein.

Brief History of RB Research

The retinoblastoma gene product (RB) was identified as a suppressor of tumor formation because it was found to be absent or mutated in many human tumors (see review by Weinberg, 1991).[1] Subsequently, findings from studies involving oncogenic tumor viruses greatly influenced the way in which research on the biological role of RB developed, favoring some aspects and occulting or neglecting others.

Indeed, a major advance in the search for RB function was the finding that the RB protein, in its hypophosphorylated form, is a target for oncoproteins encoded by the small DNA tumor viruses – E1A,[2] SV40 large T antigen (LT),[3] and E7.[4] Most intriguingly, formation of complexes with RB requires a sequence in the viral proteins that is essential for their oncogenic activity. Point mutations of the viral proteins that inhibit their oncogenic capacity also affect their binding to RB. The observed correlation between the oncogenic capacity of these viral proteins and their ability to inactivate RB, together with the fact that the oncoprotein-binding region of RB is almost invariably affected in mutant RB proteins isolated from human tumor

*Corresponding Author: Chantal E. Crémisi—Laboratoire Oncogènese, Différenciation et Transduction du Signal, CNRS UPR 9079, Institut Fédératif André Lwoff, 94801 Villejuif, France. Email: cremisi@vjf.cnrs.fr

Rb and Tumorigenesis, edited by Maurizio Fanciulli. ©2006 Eurekah.com and Springer Business+Science Media.

samples, led to the hypothesis that RB inactivation by viruses mimics natural RB mutations occurring in cancers. As a corollary, the loss of oncogenic potential that is observed when these viral oncoproteins are mutated would be directly attributable to the loss of RB-binding capacity, i.e., to their failure to inactivate RB.

Concomitantly, the E1A-activated transcription factor E2F was discovered and was suspected of binding to a number of cellular promoters, including some that are involved in regulating the G1/S transition.[5,6] The E1A effect on E2F action was found to correlate with the appearance of a free form of E2F,[7] and it was thought that RB, a phosphoprotein whose phosphorylation state is tightly regulated during the cell cycle, might be involved in this regulation. Later, E1A mutants unable to bind RB were shown to be incapable of activating the viral E2 promoter via E2F,[8] and finally, E2F was found to be complexed to hypophosphorylated RB[9] and inactivated by this interaction.[10]

To take into account all these observations, Nevins (1992)[11] proposed a model where E2F activity would be regulated by its interaction with RB, and the liberation of E2F by the viral oncoproteins would be a key event responsible for both their oncogenic properties and cell cycle progression and proliferation.

Thus, early on, RB was established as a negative transcriptional regulator, and since then a plethora of excellent studies have addressed the role of RB in the inhibition of cell-cycle progression via its interaction with E2F (see review by Dyson 1998).[12]

Over the years many prestigious reviews have taken this hypothesis for granted, and it is probably no exaggeration to say that the model has attained the status of dogma, arguably slowing down research on other functions of RB, and so delaying the comprehensive understanding of RB functions in their entirety. Indeed, the role of RB in the maintenance and induction of a differentiated phenotype, in cell survival and in senescence took a lot of time to gain widespread acceptance. The notion that RB plays a role in establishing a differentiation program and that this function might also be involved in tumor suppression was only gradually accepted, over a period of more than 10 years. To illustrate this point, I would like to mention just a study published in 2000,[13] which recognized very late, several years after the initial study[14] the role of RB in adipocyte differentiation.

Nevertheless, in 1998, Dyson[11] in his excellent review made several important points, emphasizing, for example, the fact that "pRB/E2F complexes represent only a minor fraction of the total E2F complexes and of pRB present in cell extracts. The interaction between pRB and E2F needs to be placed carefully into context. Biochemical studies have suggested that E2F is only one of many pRB targets and, to date, at least 110 cellular proteins have been reported to associate with RB. Such studies illustrate that E2F-regulation is only one aspect of pRB function and emphasize the need to identify other pRB-binding proteins that are important for tumor suppression".

In Vivo Studies

The first data clearly indicating that RB plays an important role in both differentiation and in cell survival come from three papers published in 1992 analyzing RB knockout mice.[15,16,17] The RB scientific community was surprised by the results of these studies, because at the time no one suspected that RB might have other functions besides the regulation of cell-cycle progression.

Mice deficient for RB are nonviable and die in utero between days 13.5 and 15.5 of gestation. They show defects in neurogenesis and hematopoiesis: massive cell death occurs in maturing neuronal centers, suggesting that RB is necessary for proper completion of the neuronal differentiation program. Both the neuronal and hematopoietic phenotypes of RB knockout mice are consistent with a role for RB in controlling the proliferation/differentiation of a few specific cell lineages during embryogenesis. Unfortunately, death occurs too early to indicate whether epithelial differentiation is normal in these mice.

A different type of study published the same year provided additional support for this notion: in analyzing the expression pattern of RB protein in SCID mouse fetuses by immunofluorescence and confocal laser scanning microscopy, Szekely et al (1992)[18] found that, although it was generally believed that RB protein is ubiquitously expressed, in fact there is a high degree of heterogeneity among tissues. In addition to hematopoietic and neuronal cells, high RB expression was found in various epithelial tissues such as kidney collecting tubules, teeth, skin, the mucous membranes of the digestive tract and the basal layer of stratified squamous epithelia, where it is confined in the more differentiated layer. Different epithelia showed the same characteristic RB distribution.

These results are in general agreement with those previously obtained by Bernards et al (1989)[19] for RB mRNA expression in mouse embryos and in adults. They found that RB expression was quite variable. Importantly, all these in vivo studies reveal that the role of RB family proteins will vary depending on whether the cells are proliferating or differentiating, and depending on the specific cell type. Thus, the differentiation-dependent expression of RB strongly suggests that RB may play specific roles in several tissues, and provide a starting point from which to begin to unravel the tissue-specific oncogenic effects of RB loss.

Ex Vivo Studies Involving Epithelial Cells

After these in vivo studies, cell cultures were used for detailed analysis of the molecular mechanisms used by RB during the process of differentiation.

In one study,[20] keratinocytes (HaCaT cell line and primary keratinocytes) were used as a model system to study the participation of the transcriptional activity of RB in epithelial cells. Hypophosphorylated RB was found to bind the transcriptional factor c-jun both in vitro and ex vivo, but only during the early G1 phase in proliferating cells and, importantly, during differentiation of primary keratinocytes. This interaction stimulates RB binding to the AP-1 consensus site and up-regulates c-jun transcriptional activity, as tested on collagenase and c-jun promoters. The interaction between RB and c-jun involves the small B pocket of the former (AA 612-657) and the C-terminal domain and leucine zipper region of the latter. One can conclude from this study that RB acts as a positive transcriptional regulator during epithelial keratinocyte differentiation, by interacting with a specific transcription factor (AP-1) and increasing its activity. In this context, it is of interest to keep in mind that c-jun is up-regulated when normal keratinocytes are induced to differentiate, and that both RB and c-jun have been individually localized to the upper layer of the epithelium.[18]

It is essential to determine the molecular mechanisms of interaction between viral oncoproteins and RB when the latter acts as a positive transcriptional regulator, as they might well differ from those used when RB acts as a negative transcriptional regulator. In a first approach to elucidating this point, Nead et al,[20] used transfection experiments to analyze the effect of HPV-16 E7 on RB-mediated transactivation of the c-jun promoter, and found that this transactivation is inhibited by wild type E7 but not by a mutant, E7-gly24, that retains c-jun binding but cannot bind RB. Interestingly, in the presence of the E7-gly24 mutant and RB, c-jun activation is significantly higher (60-70%) than with RB alone. Although it is not immediately obvious how to reconcile all these results with those of a previous paper, emanating from the same team,[21] which indicated that wild type HPV16-E7 transactivates c-jun, it is worth noting that the earlier report also showed that the E7-gly24 mutant activates c-jun even more than the wild type E7. This suggests that the increased activation of c-jun observed in the presence of the E7-gly24 mutant may be physiologically significant, since it is reproducible. Taken together, the results presented in these two papers[20,21] indicate that E7 partially inhibits RB-mediated activation of c-jun and can also activate c-jun independently of RB; but they do not elucidate the precise mechanism involved in viral oncoprotein inhibition of RB-mediated activation in epithelial cells.

A completely independent series of studies used the canine epithelial cell line MDCK, derived from kidney distal tubules, to characterise the functions of RB in epithelial cells; these

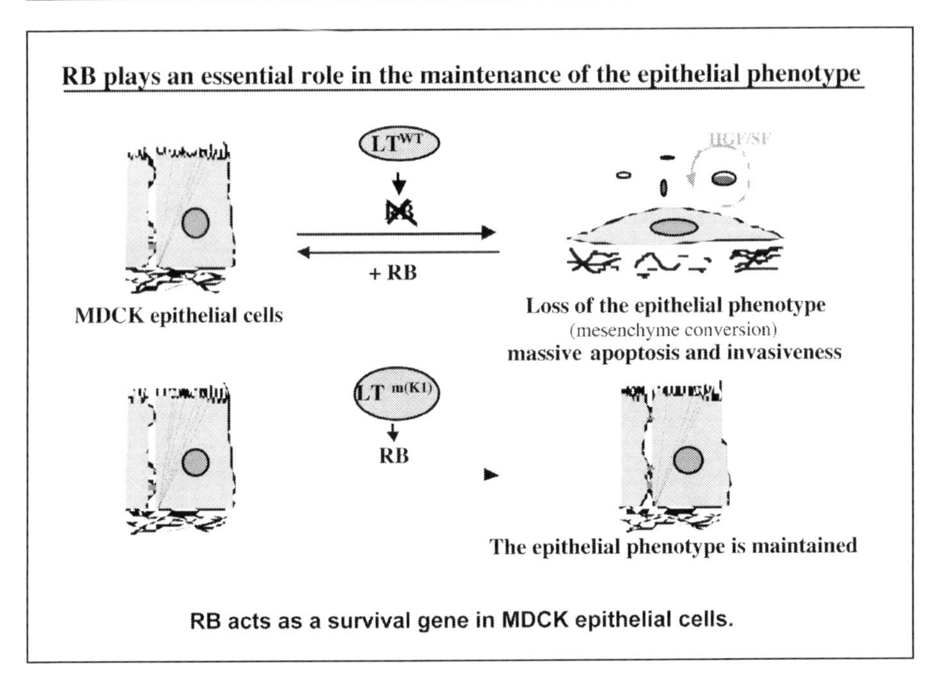

Figure 1. Inactivation of RB protein family in MDCK cells induces a complete mesenchyme conversion with a loss of expression of epithelial markers.

cells, in which p53 is not mutated, are nontumorigenic. Wild type SV40 LT and a mutant (K1) unable to bind to RB but still inactivating p53 were used to transform MDCK epithelial cells in an attempt to determine the specific effects of RB-protein family inactivation.[22,23] RB inactivation was found to induce a complete mesenchyme-like conversion, i.e., a loss of epithelial differentiation, characterised by a loss of polarity accompanied by a loss of expression of epithelial marker genes (E-cadherin, cytokeratins, desmoplakin), and an inhibition of c-fos, c-myc and p21 expression as well as that of the transcription factor vHNF1, whose expression is restricted to epithelial tissues. These changes were observed to be concomitant with the appearance of invasiveness of MDCK (LT) cells in vitro, induction of metalloproteases, and creation of a hepatocyte growth factor/scatter factor (HGF/SF) autocrine loop. In addition, a massive apoptosis occurred that was independent of the presence of serum in the culture medium.

In contrast, none of these changes were observed when the MDCK cells were transformed with the K1 mutant, i.e., when RB remained active but p53 was inactivated. This strongly suggests that inactivation of the RB-protein family is somehow responsible for the loss of epithelial phenotype upon transformation by the wild type LT.

To further confirm the role of RB, it was reexpressed in MDCK (LT) cells by RB gene transfer. This reexpression of RB partially restored the epithelial phenotype, with cell polarity and the expression of all epithelial markers as well as reexpression of c-myc and c-fos (see Fig. 1).

These studies[22,23] were among the first to demonstrate that RB plays an essential role in the maintenance of the epithelial phenotype, providing a mechanistic explanation for the highly tissue-specific RB expression pattern seen during embryogenesis.[18] An additional important point arising from this work was the observation of a correlation between the maintenance of the epithelial phenotype and cell viability: loss of the former results in loss of the latter and vice versa. Thus RB may be characterized as being a survival gene, this property being strongly linked to its function in cell differentiation. Indeed, these two properties may very likely represent two aspects of the same function.

In this experimental system, RB-protein-family inactivation was found to result in a mesenchyme conversion with creation of an auto-loop of HGF/SF accompanied by invasiveness. It is important to remember that mesenchyme conversion occurs in pathological conditions, in invasive cancers when metastases are produced – but also during embryogenic development, where it is essential in the formation of many organs, among them kidney, ovary, and mammary gland.

Determination of the Molecular Mechanisms of RB Action by Identifying Its Target Genes in Epithelial Cells

E-Cadherin Gene

The next essential question was to find the RB-target genes involved in this mesenchyme conversion and, more specifically, to determine the molecular mechanisms of RB action.

The E-cadherin gene was the first target gene to be identified.[24] E-cadherin is essential for the maintenance and function of the epithelial cell layer and also plays a pivotal role very early in development, during the compaction process of the preimplantation embryo, i.e., in the biogenesis of the epithelium.[25]

E-cadherin expression is down-regulated in tumor progression. In carcinomas, this down-regulation is associated with invasiveness, dedifferentiation and metastasis of carcinoma cells in vivo.[26] The reexpression of E-cadherin in these cells decreases their invasiveness. E-cadherin is therefore considered to be a tumor suppressor.[27]

RB was found to specifically activate transcription of the E-cadherin promoter in epithelial cells, but not in NIH3T3 mesenchymal cells.[24] This transcriptional activity is mediated by the transcription factor AP-2. Physical interaction between RB and AP-2 is observed both in vitro and in MDCK and HaCaT keratinocyte cells. It involves the N terminal domain of AP-2 and the oncoprotein binding domain and C-terminal domain of RB. Furthermore, the RB/AP-2-mediated transcriptional activation is not restricted to the E-cadherin promoter, but is also observed with the mouse cytokeratin (endoA) promoter.[24] Thus RB acts as an activator of AP-2 in epithelial cells, as it does in the case of c-jun in keratinocytes.

The fact that RB activates the expression of the E-cadherin gene, the master gene of the epithelial phenotype, definitively demonstrates that RB plays a major role in the maintenance of this phenotype. Furthermore, the AP-2 transcription factor is a cell-type-specific factor involved in the expression of many epithelial markers.[28] AP-2 has also been shown to restrict the growth of some tumor cells, so it may have some of the properties of a tumor suppressor.[29] Thus the RB anti-oncogene activates the expression of a tumor suppressor (E-cadherin) in a cell-type-specific manner through a transcription factor (AP-2) that may itself be a tumor suppressor.[24]

p21 Gene

Two other target genes of RB have been identified in epithelial cells: the p21 gene[30] and the survival gene Bcl-2.[31] The functions of these two target genes are consistent with the role of RB in differentiation and cell viability.

The p21 gene, besides its role in cell cycle regulation, has been implicated in terminal-differentiation-associated growth arrest independently of p53. Its expression is increased during several differentiation processes including that of keratinocytes.[32] During embryogenesis, the p21 expression pattern, like that of RB, correlates with terminal differentiation in several cell lineages, for example, in nasal epithelium and the outermost layer of the embryonic epidermis.[33] In adults, p21 is expressed in a highly selective manner and is found in large amounts in the fully differentiated columna. The overlapping pattern of expression between RB and p21 during embryogenesis and in adults is striking.

RB transcriptionally upregulates the expression of the endogenous p21 gene, specifically in epithelial cells, through Sp1 and Sp3 transcription factors.[30] The RB-Sp1-mediated activation of p21 is interesting because, in vivo, Sp1 family members are expressed in a tissue-specific

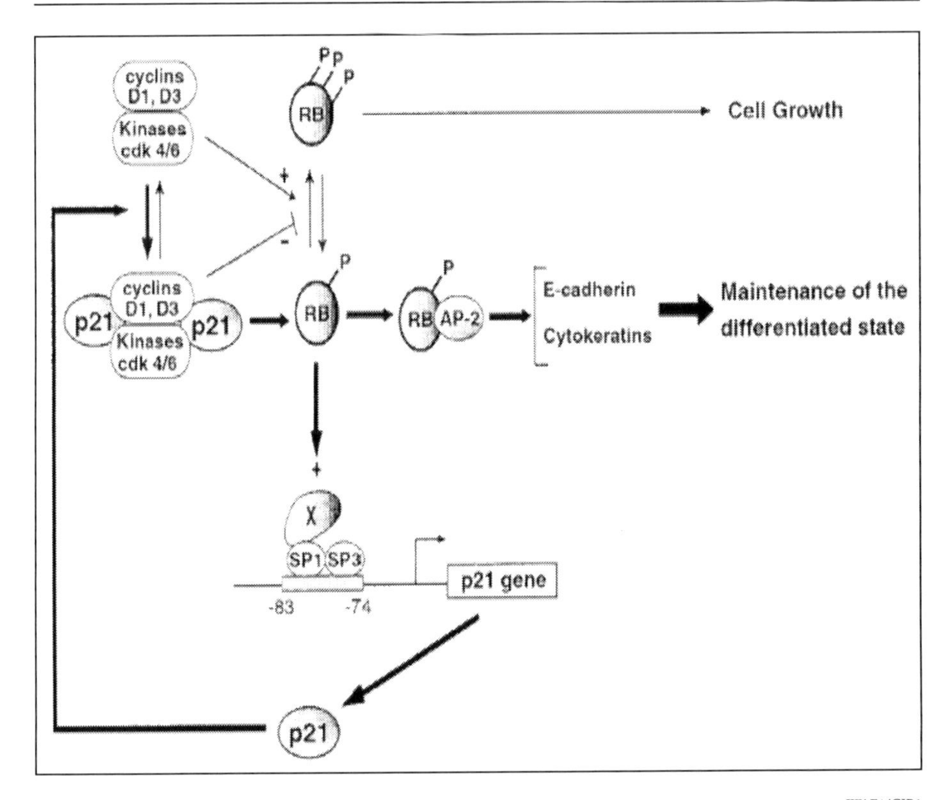

Figure 2. Hypothetical working model representing the auto-loop of regulation between RB and p21$^{WAF1/CIP1}$ in differentiated cells. RB up regulates transcriptional p21 expression through Sp1 and other as yet unidentified factor which in turn keeps RB in a hypophosphorylated state. Hypophosphorylated RB was shown to activate differentiation transcription factors such as AP-2. Finally, expression of the differentiated markers such E-cadherin and cytokeratins would contribute to the maintenance of the differentiated state. Reprinted with permission from: Decesse J et al. Oncogene 2001; 20:962-971.

manner.[34] In fact Sp1, like AP-2, is involved in the expression of many specific epithelial markers, including E-cadherin and several keratin genes.[28,35] In many of these specific epithelial promoters, and in particular, in the p21 promoter,[29] SP1 binding sites were found to be in close proximity to AP-2 sites.

Sp1 family factors were previously described to be involved in RB-mediated activation of several promoters (c-fos, c-myc, TGFβ-1 and IGFII). In all those cases, the interaction between RB and Sp-1 was found to be indirect. It was suspected that p300 might mediate the interaction between RB and Sp-1 since, in epithelial cells, p300 is able to bind to the RB target sequence on the p21 promoter and, furthermore, to interact with Sp-1.[36]

The results of Decesse[30] bring to light an auto-loop of regulation between RB and p21 that is essential for maintenance of the epithelial differentiation phenotype (see Fig. 2). In vivo during epithelial differentiation, the expression of both genes is increased, as already mentioned. Increased p21 levels would maintain RB in a hypophosphorylated state, allowing it to interact with specific differentiation-associated transcription factors such as AP-2 or Sp1 family members, resulting in the activation of differentiation-specific genes such as E-cadherin, cytokeratin, p21 and Bcl-2. Only the hypophosphorylated form of RB is present during differentiation and could be found complexed with the above-mentioned factors. These specific protein interactions could in turn allow RB to act as a positive transcriptional regulator, thereby maintaining the differentiated state.

Bcl-2 Gene: In Live Cells, RB Is Associated with Chromatin of the Target-Gene Promoters

The developmental pattern of Bcl-2 expression suggests that Bcl-2 has a role beyond the regulation of cell death. Bcl-2 expression is prominent in the nervous system and during kidney development, and interestingly, Bcl-2 expression levels do not mirror patterns of cell death in all tissues: changes in its expression match cell differentiation more closely than cell death.[37] Indeed, beside its function in cell survival, Bcl-2 has now been shown to intervene in several differentiation systems, including epithelial differentiation.[38] Thus Bcl-2 is regulated in both a tissue-specific and temporal manner.

RB specifically activates transcription of Bcl-2 in epithelial cells but not in NIH3T3 mesenchymal cells.[31] This RB-mediated activation of Bcl-2 is independent of p53, as is the apoptosis mediated by RB inactivation,[22] suggesting a direct molecular mechanism by which RB might inhibit apoptosis independently of p53. Furthermore, this RB-mediated activation of Bcl-2 in epithelial cells also requires interaction with the AP-2 transcription factor,[31] as previously demonstrated for the activation of the E-cadherin gene by RB in these same cells.[24] By monitoring protein-DNA interactions in living cells using formaldehyde cross-linking and chromatin immunoprecipitation (ChIP) followed by quantitative PCR using a LightCycler (Roche), it was shown that endogenous hypophosphorylated RB and AP-2 bind to the same Bcl-2 promoter sequence.[31] Histone H4 acetylation at this site was also monitored, and RB target sequences were found to be highly acetylated, consistent with active gene transcription at the site. In addition, RB and AP-2 were shown to bind to the E-cadherin gene promoter in vivo, consistent with regulation of this promoter by both AP-2 and RB in epithelial cells.

These results indicate that RB is recruited to the native Bcl-2 and E-cadherin gene promoters by AP-2. More importantly, they provide the first direct experimental evidence that the molecular mechanism used by RB, when acting as a positive regulator, involves direct activation of the target gene, rather than the downregulation of a repressor. The fact that this RB-mediated activation is correlated with increased histone acetylation suggests the recruitment of a HAT to the complex. This mechanism might in fact be general since, in another model, it has been shown by ChIP that RB is recruited by CBFA1 to specific differentiation-associated promoters during osteogenic differentiation.[39]

Thus, we are faced with two different molecular mechanisms, depending whether RB is acting as a negative transcriptional regulator, when inhibiting cell growth, or as a positive transcriptional regulator, when promoting cell differentiation and cell survival. In the first case, it is accepted that RB represses E2F family proteins and might recruit an HDAC to the complex. However, it has to be mentioned that, using ChIP, several groups could not find RB binding to transcriptionally repressed genes in living cells in G0 as expected.[40-42] In the second case, RB is recruited to differentiation and survival gene promoters by specific transcription factors, and enhances their transcriptional activity. Very likely other factors are also associated with this protein complex, including HATs, since the Bcl-2 and E-cadherin promoters were found to be specifically acetylated when RB and AP-2 bound to them (see model in Fig. 3).

Interaction between Viral Oncoproteins and RB in Epithelial Cells

An important feature of RB biology is the interaction of the nucleoprotein complex RB/transcription factor and chromatin with the viral oncoproteins, which are known to inactivate RB.

When RB acts as a negative transcriptional regulator, it inhibits the E2F family transcription factors by binding to them. The presence of an oncoprotein in the cell dissociates the interaction of RB from E2F, which allows E2F to recover its activity. In epithelial cells, when RB acts as a positive transcriptional regulator, it binds to and activates AP-2. The viral oncoprotein LT inhibits AP-2 activity by binding to the complex RB/AP-2 without dissociating it.[24] It has been clearly demonstrated that a tripartite complex between SV40 LT, RB and AP-2 is present in living cells, and that the physical and functional interaction between LT and AP-2

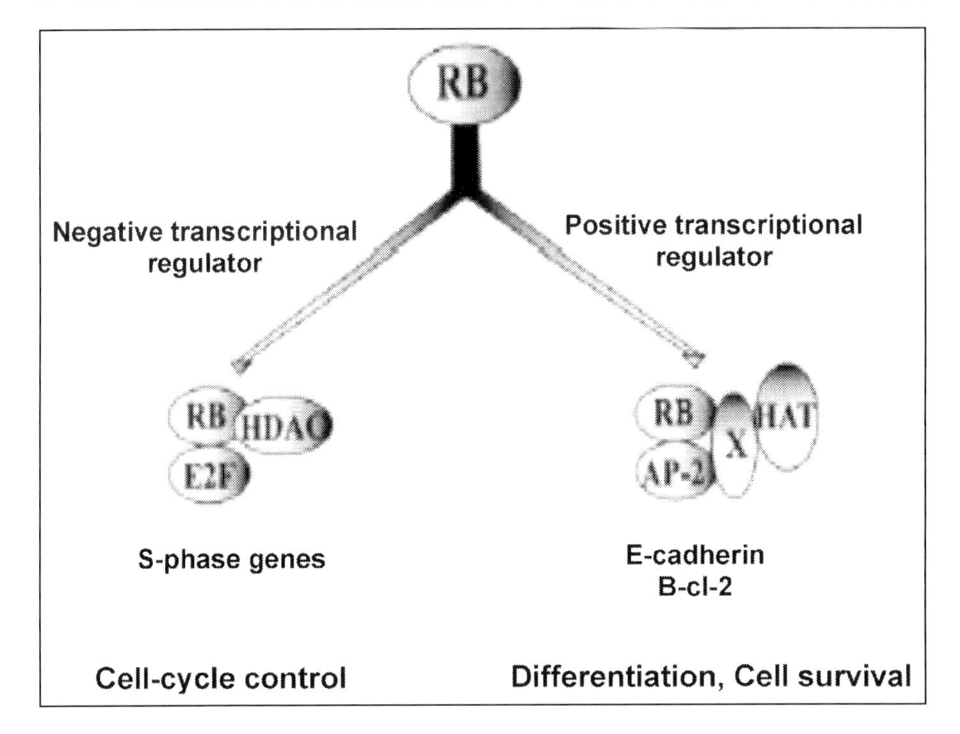

Figure 3. Model for RB function. Reprinted with permission from: Decary S et al. Mol Cell Biol 2002; 22:7877-7888.

is mediated by RB.[24] To our knowledge, this is only the second example studied in detail – after the RB/E2F complex, which was studied mainly in fibroblasts—where a viral oncoprotein has been shown to affect the activity of a transcription factor through RB. In the present case, in contrast to the first one, the consequence of inhibiting a transcriptional activity results in the loss of a differentiated phenotype.

The inhibition by LT of the RB-mediated activation of AP-2 activity may constitute an important mechanism through which the oncoprotein exerts its oncogenic effect.[24] Indeed, the AP-2 transcription factor is particularly involved in epithelial gene expression,[43] and the small DNA tumor viruses producing these oncoproteins all show varying degrees of epithelial tropism. Moreover, 90% of all human tumors originate from epithelial cells. Therefore, it was hypothesized[24] that the LT-RB complex inactivating AP-2 might play an important role during the dedifferentiation processes occurring during tumor progression (see Fig. 4), and that the overall transcription of genes containing an AP-2 binding site may be regulated by RB. One candidate gene is that of AP-2 itself, which is positively autoregulated by its own product.[44]

An important question remaining to be elucidated is whether LT dissociates the RB/AP-2 complex from chromatin or, alternatively, if LT binds to this complex on chromatin and inhibits its activity in situ, e.g., by inhibiting the recruitment of other factors or cofactors required for the transcriptional activity.

RB is a tumor suppressor having several functions in cell-cycle regulation, differentiation, cell survival and senescence. How many of these functions are involved in tumor-suppression? The majority of tumor-derived RB mutants described are defective for all of RB's biochemical activities, making it difficult to discern the relative contribution of these activities to RB-mediated tumor suppression, possibly suggesting that in vivo all these activities of RB are indissociably intertwined.

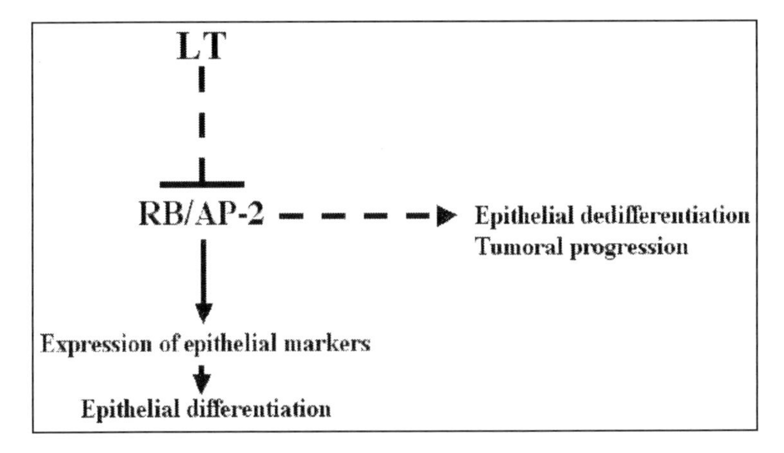

Figure 4. LT, RB and AP-2 interaction in epithelial cells.

References

1. Weinberg R. Tumor suppressor genes. Science 1991; 254:1138-1146.
2. Whyte P, Buchkovich K, Horowitz J et al. Association between an oncogene and a antioncogene: The adenovirus E1A proteins bind to the retinoblastoma gene product. Nature 1988; 334:124-129.
3. DeCaprio J, Ludlow M, Figurege J et al. SV40 large tumour antigen forms a specific complex with the product of the retinoblastoma susceptibility gene. Cell 1988; 54:275-283.
4. Dyson N, Howley P, Munger K et al. The human papilloma virus-16 E7 oncoprotein is able to bindto the retinoblastoma gene product. Science 1989; 243:934-937.
5. Kovesdi I, Reichel R, Nevins J. Identification of a cellular transcription factor involved in E1A trans-activation. Cell 1986; 45:219-228.
6. Mudryj M, Devoto S, Hiebert W et al. A role for the adenovirus inducible E2F transcription factor in a proliferation dependent signal transduction pathway. EMBO J 1990; 9:2179-2184.
7. Bagchi S, Raychaudhuri P, Nevins JR. Adenovirus E1A proteins can dissociate heterodimeric complexes involving the E2F transcription factor: A novel mechanism for E1A trans-activation. Cell 1990; 62:659-669.
8. Chellappan S, Kraus V, Kroger B et al. Adenovirus E1A, simian virus 40 tumor antigen, and human papillomavirus E7 protein share the capacity to disrupt the interaction between transcription factor E2F and the retinoblastoma gene product. Proc Natl Acad Sci USA 1992; 89:4549-4553.
9. Chellapan S, Hiebert S, Mudryij M et al. The E2F transcription factor is a cellular target for the RB protein. Cell 1991; 65:1053-1061.
10. Hiebert S, Chellappan S, Horowitz J et al. The interaction of RB with E2F coincides with an inhibition of the transcriptional activity of E2F. Genes Dev 1992; 6:177-185.
11. Nevins J. E2F: A link between the Rb tumor suppressor protein and viral oncoproteins. Science 1992; 258:424-429.
12. Dyson N. The regulation of E2F by pRB-family proteins. Genes Dev 1998; 12:2245-62.
13. Classon M, Kennedy B, Mulloy R et al. Opposing roles of pRB and p107 in adipocyte differentiation. Proc Natl Acad Sci USA 2000; 97:10826-10831.
14. Chen L, Riley J, Chen Y et al. Ritinoblastoma protein positively regulates terminal adipocyte differenciation through direct interaction with C/EBPα. Genes Dev 1996; 10:2794-2804.
15. Clarke A, Maandag E, van Roon M et al. Requirement for a functional Rb-1 gene in murine development. Nature 1992; 359:328-331.
16. Jacks T, Fazeli A, Schmitt E et al. Effects of an Rb mutation in the mouse. Nature 1992; 359:295-300.
17. Lee E, Chang C, Hu N et al. Mice deficient for Rb are nonviable and show defects in neurogenesis and haematopoiesis. Nature 1992; 359:288-94.
18. Sezkely L, Jiang W, Buclic-Jakus F et al. Cell type and differentiation dependent heterogeneity in retinoblastoma protein expression in SCID mouse fetuses. Cell Growth Diff 1992; 3:149-156.
19. Bernards R, Schakleford M, Gerber R et al. Structure and expression of the murine retinoblastoma gene and characterization of its encoded protein. Proc Natl Acad Sci USA 1989; 86:6474-6478.
20. Nead M, Baglia L, Antinore M et al. Rb binds c-Jun and activates transcription. Embo J 1998; 17:2342-52.

21. Antimore M, Birrer M, Patel D et al. The human papillomavirus type 16 E7 gene product interacts with and trans-activates the AP1 family of transcription factors. EMBO J 1998; 15:1950-1960.
22. Martel C, Batsché EF, Harper F et al. Inactivation of the retinoblastoma gene product or an RB-related protein by SV40 T antigen in MDCK epithelial cells results in massive apoptosis. Cell Death Diff 1996; 3:61-74.
23. Martel C, Harper F, Cereghini S et al. Inactivation of retinoblastoma family proteins by SV40 T antigen results in creation of a Hepatocyte growth factor/scatter factor autocrine loop associated with an epithelial-fibroblastoid conversion and invasivness. Cell Growth Diff 1997; 8:165-178.
24. Batsche E, Muchardt C, Behrens J et al. RB and c-Myc activate expression of the E-cadherin gene in epithelial cells through interaction with transcription factor AP-2. Mol Cell Biol 1998; 18:3647-3658.
25. Larue L, Ohsugi M, Hirchenhain J et al. E-cadherine null mutant embryos fail to form a trophectoderm epithelium. Proc Natl Acad Sci USA 1994; 91:8263-8267.
26. Riethmacher D, Brinkmann V, Birchmieir C. A targeted mutation in the mouse E-cadherin gene results in defective preimplantation development. Proc Natl Acad Sci USA 1995; 92:855-859.
27. Behrens J, Mareel M, Van Roy F et al. Dissecting tumor cell invasion: Epithelial cells acquire invasive properties after the loss of uvomorulin-mediated cell-cell adhesion. J Cell Biol 1989; 108:2435-2447.
28. Fuchs E. Epidermal differentiation and keratin gene expression. J Cell Sci 1993; 17:197-208.
29. Zeng Y, Somasundaram K, el-Deiry WS. AP-2 inhibits cancer cell growth and activates $p21^{WAF1/CIP}$. Nat Genet 1997; 15:78-82.
30. Decesse J, Medjkane S, Datto M et al. RB regulates transcription of the p21/WAF1/CIP1 gene. Oncogene 2001; 20:962-71.
31. Decary S, Decesse J, Ogryzko V et al. The Retinoblastoma protein binds the promoter of the survival gene Bcl2 and regulates its transcription in epithelial cells through transcription factor AP-2. Mol Cell Biol 2002; 22:7877-7888.
32. Missero C, Calautti E, Eckner R et al. Involvement of the cell inhibitor Cip1/WAF1 and the E1a-associated p300 protein in terminal differentiation. Proc Natl Acad Sci USA 1995; 92:5451-5455.
33. Parker S, Eichele G, Zhang P et al. p53-independent expression of p21Cip1 in muscle and other terminally cells. Science 1995; 267:1024-1027.
34. Saffer J, Jackson P, Annarella M. Developmental expression of Spl in the mouse. Mol Cell Biol 1991; 11:2189-2199.
35. Behrens J, Lowrick O, Klein-Hitpass L et al. The E-cadherin promoter: Functional analysis of a G.C-rich region and an epithelial cell-specific palindromic regulatory element. Proc Natl Acad Sci USA 1991; 88:11495-9.
36. Billon N, Carlisi D, Datto MB et al. Cooperation of Sp1 and p300 in the induction of the CDK inhibitor p21WAF1/CIP1 during NGF-mediated neuronal differentiation. Oncogene 1999; 18:2872-2882.
37. Veis D, Sorenson C, Shutter J et al. Bcl-2-deficient mice demonstrate fulminant lymphoid apoptosis, polycystickidneys, and hypopigmented hair. Cell 1993; 75:229-240.
38. Lu P, Lu L, Rughetti A et al. bcl-2 overexpression inhibits cell death and promotes the morphogenesis, but not tumorigenesis of human mammary epithelial cells. J Cell Biol 1995; 129:1363-78.
39. Thomas D, Carty S, Piscopo D et al. The retinoblastoma protein acts as a transcriptional coactivator required for osteogenic differentiation. Mol Cell 2001; 8:303-16.
40. Takahashi Y, Rayman J, Dynlacht D. Analysis of promoter binding by the E2F and pRB families in vivo: Distinct E2F proteins mediate activation and repression. Genes Dev 2000; 14:804-816.
41. Wells J, Boyd E, Fry C et al. Target gene specificity of E2F and pocket protein family members in living cells. Mol Cell Biol 2000; 20:5797-807.
42. Rayman J, Takahashi Y, IndjeinV et al. E2F mediates cell cycle-dependent transcriptional repression in vivo by recruitment of on HDAC1: mSin3B corepressor complex. Genes Dev 2002; 16:933-947.
43. Leask A, Byrne C, Fuchs E. Transcription factor AP-2 and its role in epidermal-specific gene expression. Proc Natl Acad Sci USA 1991; 88:7948-7952.
44. Bauer R, Imhof A, Pscherer A et al. The genomic structure of the human AP-2 transcription factor. Nucleic Acids Res 1994; 22:1413-1420.

CHAPTER 2

pRb in the Differentiation of Normal and Neoplastic Cells

Deborah Pajalunga, Grazia Camarda and Marco Crescenzi*

Abstract

This chapter deals with the role played by the retinoblastoma protein (pRb) in a variety of differentiation processes. After broadly reviewing the current knowledge on this issue, it points at two common themes. The first is the exclusive involvement of pRb in the final maturation stages of each lineage, so that the functional ablation of the protein produces relatively subtle differentiation defects. The second is that, at least in the cell types more thoroughly investigated, pRb exerts its pro-differentiation potential by enhancing the activities of transcription factors that are key regulators of tissue-specific differentiation.

Finally, the hypothesis is put forward that pRb plays a role in the final differentiation stages of a much wider range of cell types than currently recognized. It is proposed that one reason for the well-know, poorly-understood, inverse relationship between differentiation and malignancy is the functional impairment of pRb and possibly its family members in the vast majority of human cancers.

Introduction

Tumor cells are uniformly characterized by unchecked proliferation on one side and impaired or arrested differentiation on the other.[1] In general, the degree of differentiation of tumors correlates inversely with their malignancy, the more undifferentiated neoplasias being also the more aggressive. The mechanisms through which altered proliferation and impaired differentiation are linked are only partially understood.

The general rule that all malignant tumors show altered differentiation applies even to malignancies ostensibly comprised of highly or even terminally differentiated cells. A case in point is that of multiple myeloma. In this disease, the vast majority of tumor cells are terminally differentiated plasma cells, while the malignancy grows through the expansion of a minor compartment of cells whose differentiation is arrested at a stage compatible with proliferation.[2] Conversely differentiation, when allowed to proceed in a tumor cell, can take over cell cycle control and bring the neoplastic cell to a halt. It is well known that differentiation tends to oppose cellular transformation. In extreme but not uncommon cases, terminal differentiation coexists with malignant transformation. Although terminally differentiated cells are unable to proliferate by definition and cannot possibly be transformed, spontaneous terminal differentiation often takes place even in highly aggressive cancers in a fraction–sometimes the majority–of the tumor cells. Such cells cease proliferating, thus demonstrating that differentiation can suppress the transforming events that lead to tumorigenesis. By way of example, in the

*Corresponding Author: Marco Crescenzi—Department of Environment and Primary Prevention, Istituto Superiore di Sanitá, Rome, Italy. Email: crescenz@iss.it

Rb and Tumorigenesis, edited by Maurizio Fanciulli. ©2006 Eurekah.com and Springer Business+Science Media.

solid tumor rhabdomyosarcoma, most cells are often undifferentiated and morphologically uncharacteristic. However, a variable percentage of cells derived from the malignant clone terminally differentiate into muscle fibers.[3] Understanding the molecular underpinnings of the conflict between neoplastic transformation and differentiation is a fundamental question in cancer biology. As it is true of all basic questions in cancer, finding answers should provide us with potential targets for therapeutic interventions. Indeed, it should be stressed that differentiation of some tumors can also be elicited both in vitro and in vivo by chemical treatments. We must understand the molecular mechanisms through which differentiation can force a tumor cell to stop as in the instances of differentiation therapies for myeloid leukemias and neuroblastomas.

The universal character of the transformation/differentiation antithesis suggests that one or very few mechanisms underlay it. While it is known that a large number of genetic and epigenetic alterations concur to cell transformation in a seemingly endless variety of combinations, we suggest that most or all transforming mechanisms share a common theme entailing impaired differentiation. Specifically, we hypothesize that one fundamental reason for the inverse correlation between differentiation and malignancy is the near–universal inactivation of the retinoblastoma protein (pRb) pathway in human tumors. Such inactivation alters the control of the cell cycle and contributes to determine the unchecked proliferation of tumor cells. At the same time, we contend, it impairs cell differentiation in a much wider variety of cell types than currently appreciated. Impaired differentiation is far from being a marginal byproduct of Rb pathway inactivation. On the contrary, it is a necessary condition for sustained proliferation of tumor cells in those cases in which their normal counterparts terminally differentiate into nonproliferating or postmitotic cells. This is very frequently true, as for example in the cases of most epithelial and hematopoietic cells which are almost always postmitotic in the final stages of their differentiation. Even when the differentiation of a given cell type is nonterminal (e.g., hepatocytes, thyrocytes), it is still accompanied by very low proliferation rates which are hardly compatible with neoplastic transformation.

The pRb Pathway in Normal and Neoplastic Cells

The tumor suppressor protein pRb is a central regulator of cell homeostasis, involved in the control of such critical functions as proliferation, differentiation, and programmed cell death. In the cell cycle, pRb exerts its activity in close proximity to the restriction point, regulating the decision to enter S phase. In its cell cycle regulatory capacity, pRb is primarily regulated through phosphorylation. Un- or hypo-phosphorylated pRb is conventionally regarded as "active" and prevents entry into S phase. During G1 phase of an unperturbed cell cycle, pRb is progressively phosphorylated by the cyclin D-dependent kinases cdk4 and cdk6 and the cyclin E/cyclin A-dependent cdk2. Phosphorylation of pRb "inactivates" it, thereby allowing advancement into S phase. pRb phosphorylation allows cell cycle progression mainly by releasing transcription factors of the E2F family. The E2F factors, when bound by hypophosphorylated pRb, form complexes that bind target promoters bearing E2F binding sites and actively repress transcription. Upon phosphorylation, pRb releases the E2F factors, that promote transcription of a large number of genes, many of which are essential regulators or direct effectors of DNA replication. The kinases that phosphorylate pRb are controlled by a variety of mechanisms at different levels. One prominent regulation is exerted by two groups of inhibitory molecules, the INK4 and KIP families. The INK4 family consists of four members, commonly indicated as p15, p16, p18, and p19 from their molecular weights. The INK4 inhibitors have binding specificity for the cyclin D-dependent cdk4 and cdk6 kinases and, when bound to them, prevent their forming complexes with the activating cyclins. The KIP inhibitors include p21, p27, and p57, have the ability to bind all cyclin/cdk complexes or their cyclin moieties alone, either way inhibiting cdk activity.

This highly simplified view of cell cycle regulation in G1 is summarized in Figure 1. It includes only those players whose alterations are frequently involved in pRb pathway inactivation in the course of neoplastic transformation. It is designed solely to serve this discussion,

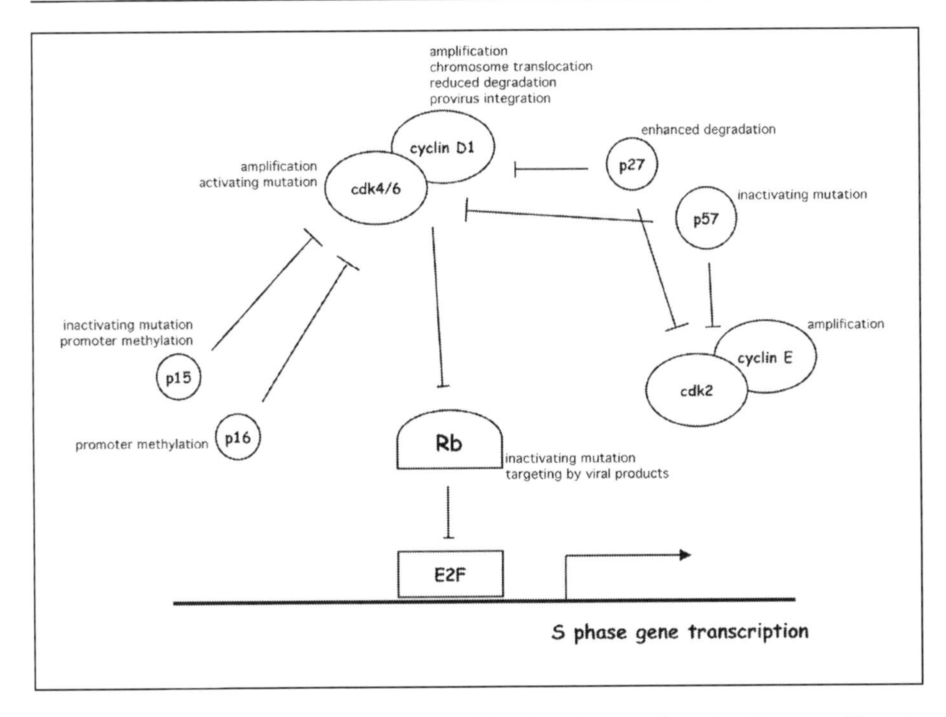

Figure 1. The pRb pathway. A simplified schematic of the pRb regulatory pathway showing some of the main factors. Mechanisms altering the function of each regulator in human cancers are indicated.

with no attempt to approximate a full description of our current understanding of G1 phase regulation.

In the last ten years it has been shown that the Rb pathway is impaired or altogether inactivated in virtually all human tumors. This remarkable discovery suggests that it is nearly impossible for a human cell to undergo neoplastic transformation without losing the restraint provided by pRb. However, pRb is directly inactivated via mutations or deletions or is targeted by viral proteins in a minority of neoplasias. Most often it is one of the components of the pRb pathway shown in Figure 1 to be altered in a number of different ways. A partial list of pRb pathway alterations in tumor cells includes reference 4 and references therein:

- amplification of the cyclin D1, -D2, -E, the cdk4, or the cdk6 gene
- chromosomal translocation of the cyclin D1 locus generating overexpression of the gene product
- activating mutations of cdk4
- inactivating mutations of the p16 or p57 cdk inhibitor
- methylation-mediated silencing of the p15 or p16 promoter
- accelerated degradation of p27
- reduced degradation of D cyclins
- activation of D cyclins by provirus integration

The main functional outcome of these alterations is constant: pRb inactivation and E2F release. Thus, the cell cycle regulatory function of pRb is suppressed in tumor cells by direct inactivation of the Rb gene or its product or by increased phosphorylation due to alterations in its regulatory pathway.

Most of the Rb-inactivating events that concur to neoplastic transformation also impinge on the two other pRb family members, p107 and p130. Whereas direct Rb gene mutations or deletions clearly affect only pRb, viral oncoproteins such as human papillomavirus E7 target all three family members. Likewise, the three pRb family proteins are targets of the same kinases[5] that presumably, when hyperactive, similarly affect all of them. Thus, current knowledge allows us to assume that the impairment of the pRb cell cycle functions is mirrored by related alterations in the capacities of its cousins.

Impairment of pRb cell cycle-related functions does not automatically translate into repression of its differentiation promoting capabilities. Indeed, these two sets of functions can be genetically separated,[6] as it has been found that some Rb mutants unable to stably bind E2F and induce cell cycle arrest can still promote differentiation. However, differentiation is tightly associated with pRb hypophosphorylation in numerous cell types.[7] In addition, naturally occurring "low risk" pRb mutants that retain differentiation-promoting capacity are mostly associated with benign retinomas,[6] suggesting again that differentiation counteracts full neoplastic transformation. To our knowledge there is no instance in which abnormal, ectopic phosphorylation of pRb is compatible with unimpaired differentiation. Thus, for the rest of the discussion, we will assume that any reduction in the ability of pRb (or its family members) to control the cell cycle is reflected in an overt or subtle modification of its differentiating properties.

pRb Is Involved in the Differentiation of a Growing Number of Cell Types

It is very instructive to summarize briefly the chronology of the discovery of the differentiation role of pRb. Rb knockout (KO) mice and their derivatives led to the conclusion that the differentiation of several cell types depends on pRb. In 1992, three groups independently reported that Rb KO mice die in utero displaying defective maturation of erythrocytes and neurons.[8-10] Two years later, closer examination revealed that in Rb KO mice lens cell differentiation was also impaired, being characterized by excess proliferation, reduced expression of differentiation markers, and apoptosis.[11] Examination of Rb KO mice brought us so far. However, the phenotype of these mice does not tell the whole story. For example, the severe impairment of skeletal muscle differentiation was first discovered in vitro in 1994[12] by examining MyoD-converted Rb KO fibroblasts and confirmed in vivo only two years later in partially rescued Rb KO mice.[13] Likewise, pRb has been shown to be essential for adipocyte differentiation in vitro,[14,15] while a fat phenotype in the KO mice has not been reported.

Subtler defects have been found in lineages for which an in vivo phenotype is not described. The differentiation of hematopoietic lineages other than the erythroid is also influenced by Rb. A role for pRb in monocyte/macrophage differentiation has been recognized as early as 1996[16] and later confirmed by direct suppression of pRb.[17,18] Some evidence exists indicating the involvement of pRb in granulocyte differentiation, too.[17,19]

Further cell types have been found to require pRb for their normal differentiation processes. In 2001 a role for pRb in osteoblast differentiation was recognized,[20] providing a molecular basis for the old epidemiological observation that osteosarcoma is one of the most common cancers arising in survivors of familial retinoblastoma.[21] Conditional Rb KO mice allowed to unveil in 2003 a previously unrecognized role of pRb in keratinocyte differentiation. Finally, in 2003, Rb KO mice were belatedly found to bear placental defects. This finding explains, at least in part, the previous recognition that some differentiation defects in erythroid and central nervous system cells are noncell-autonomous.[22,23] The proposal that at least in part such differentiation defects stem from hypoxia[24] finds a plausible mechanistic explanation in the placental abnormalities of Rb KO embryos.[25] In addition, these abnormalities suggest that yet another cell type, the placental trophoblast, might derive differentiation aberrations from the lack of pRb function.

A few cell types have been specifically reported to differentiate normally in the absence of pRb. One instance is mammary epithelium, as Rb KO cells formed histologically and functionally normal mammary glands when transplanted into wild-type female mice.[26] A second example is provided by Rb KO prostate epithelium, which has been reported to give rise to fully differentiated and morphologically normal prostate tissue when transplanted into nude mice.[27] However, as always with negative results, caution should be exerted. Very subtle differentiation defects might go undetected by histological examination and manifest themselves only in specific physiological or pathological circumstances. The presence of pRb had been deemed irrelevant for granulopoietic differentiation, a contention later disproved as discussed above. In addition, the potential role of the other pocket proteins in the differentiation of breast and prostate cells has not been investigated.

It is not easy to separate the effects of pRb on the cell cycle from those exerted on differentiation. The available data do not always allow to conclude that in a given cell type a differentiation defect is primary rather than being an indirect consequence of a perturbation of the cell cycle produced by the absence of pRb. For instance, the role of pRb in retinal rod photoreceptor differentiation might be mediated exclusively by the inability of Rb KO precursors to exit the cell cycle. However, for the examples provided above, good evidence exists suggesting an involvement of pRb in the expression of the differentiation program itself rather than exclusively in the cell cycle.

The pRb family members p107 and p130 are also involved in the differentiation processes of at least some cell types. p107/p130 double-KO mice show severe defects in limb development due to deregulated chondrocyte growth.[28] These mice also show defective keratinocyte differentiation and delayed hair follicle morphogenesis and tooth development. Whether the defects of double-KO mice should be attributed to pure cell cycle deregulation or at least partially to impairment of cell-specific differentiation programs remains to be determined. In addition, p107 and p130 single KO mice with a prevalent Balb/cJ genetic background display a variety of developmental defects.[29,30] Thus, whereas this review focuses mainly on pRb, the role of its cognate proteins should not be disregarded. The three pRb family members share regulatory mechanisms and functional properties and in many cases the specific role of each member in a biological process cannot be easily untangled from those of the others. In consequence, the reasoning here applied to pRb extends to its family proteins.

In conclusion, the differentiation of a fairly large number and broad variety of cell types is influenced by one or more of the pRb family members. We have showed that, as more and more cell types are examined closely, a differentiation role for pRb and its family proteins is recognized in far more tissues than initially suggested by the phenotype of Rb KO mice. Thus, lack of in vivo evidence for pRb differentiation activity does not necessarily rule it out. The number of cell types whose differentiation is affected by pocket proteins is still growing and we suggest that eventually most lineages will show some degree of pRb-dependence in their differentiation.

pRb-Regulated Differentiation: Common Themes

In our view, two facts recur in studies of pRb-influenced differentiation processes. First, pRb seems to be involved exclusively in the final stages of differentiation, as the development of precursor cells is generally normal numerically, morphologically, and functionally. Second, in the cases analyzed more in depth, pRb appears to exert its activities by functionally interacting with key transcription factors regulating differentiation.

The first generalization derives from the observation that in all cases in which a differentiation role for pRb has been recognized (see Table 1), differentiation defects are confined to the final cell types in each lineage, while precursor cells are ostensibly normal. For example, while skeletal Rb KO muscle fibers display impaired tissue-specific gene expression and persistent, ectopic DNA synthesis, myoblasts are seemingly normal.[12,13,31] Likewise, Rb KO erythrocyte precursors are capable of going normally through most of their differentiation pathway. A defect is present in the last differentiation stages, leading to reduced numbers of circulating erythrocytes, many of which show an immature phenotype with nucleus retention.[8-10] The contention that pRb is required for late differentiation steps in this lineage is also supported by

Table 1. Involvement of pRb in the differentiation of diverse cell types and interacting transcription factors

Cell Type	Transcription Factor Interacting with pRb	Physical Interaction	Key References
Neuron	?		8,10,24,39,40
Erythrocyte	?		8-10
Granulocyte	C/EBPε	yes	17, 19
Monocyte	NF-IL6	yes	16
Lens cell	?		11, 41
Skeletal muscle	MyoD, MEF2C	MyoD, controversial	12,31,42,43
Adipocyte	C/EBP		14,46
Osteocyte	CBFA1	yes	20
Keratinocyte	?		44
Retinal rod photoreceptor	?		45

in vitro data showing that suppression of pRb in an early culture of hematopoietic progenitor cells has no effect, while similarly treating a late culture produces a strong inhibition of erythroid colonies.[32] In an analogous fashion, central nervous system abnormalities in Rb KO mice are described as ectopic proliferation and excess apoptosis in mature neurons, but neurogenesis itself is grossly normal.[8-10,33] In the bone, pRb is dispensable for early osteoblastic differentiation, but required for the expression of such late differentiation markers as osteocalcin and for mineralization.[20] Finally, the role played by the other pocket proteins, p107 and p130, should not be forgotten. The combined absence of these two proteins severely alters limb cartilage development and late-stage chondrocyte differentiation.[28,34] The examples just cited illustrate the broad conclusion that pocket-protein deficient cell lineages are competent to initiate their differentiation programs, as indicated by their ability to express early markers, but fail to achieve a fully differentiated state.[33]

Our second generalization states that pRb acts on differentiation programs by interacting functionally, and sometimes physically, with transcription factors that are key regulators of differentiation and facilitating their activities. Table 1 reports the cell types for which a differentiation role for pRb has been established. In several cases, it has been shown that pRb enables critical transcription factors or enhances their activity, as also reported in Table 1. The means through which pRb performs these functions are far from clear. For example, pRb binds the transcription factor NF-IL6 in the course of monocytic differentiation and it has been proposed that in this specific case pRb acts like a chaperone protein, enhancing the ability NF-IL6 to bind DNA and transactivate its target genes.[16] In skeletal muscle differentiation, pRb enhances the activity of MEF2 via a poorly understood, MyoD-dependent mechanism.[35] In addition, pRb has been proposed to promote skeletal muscle differentiation by disrupting the transcription-inhibitory MyoD-HDAC1 interaction.[36] Finally, pRb has been shown to mediate the degradation of the muscle-differentiation inhibitor EID-1.[37] It is also likely that the ability of pRb to bind a variety of chromatin-modifying proteins[38] is an important, possibly a crucial factor in determining its differentiation-modulating capacities. However, this possibility has not been thoroughly investigated. Altogether, a variety of mechanisms have been proposed for pRb-mediated enhancement of differentiation and no unifying features seem to emerge yet. However, it appears reasonable to propose that pRb might enhance the activity of even more transcription factors in the examples of differentiation in which such an activity has not yet been described. Thus, we suggest that Rb-mediated enhancement of tissue-specific transcription factors should be looked for and investigated in tissue types that require pRb for optimal differentiation.

pRb-Mediated Impairment of Differentiation in Cancer?

Although neoplasias are characterized by altered or blocked differentiation, this impairment is rarely such as to prevent identification of the normal counterpart of the tumor cell. Indeed, tumor classification has long been based on purely morphological grounds. The conservation of cell-type specific morphological features and tissue-specific markers characteristic of normal cells indicates that in the neoplastic ones differentiation is only partially arrested and mostly in its final steps.

The well-known, universal impairment of cell differentiation in tumors suggests that one or a few common mechanisms exist in all neoplasias, invariably producing limited but biologically significant alterations in their differentiation processes. Loss of pocket-protein function, a characteristic of most if not all cancers, impairs the final steps in differentiation. Thus, we propose that it might explain, at least in part, the ubiquitous differentiation phenotype of tumor cells. This proposal implies that pocket proteins significantly regulate differentiation in most cells, a plausible hypothesis since more and more cell types are found to depend on pRb family proteins for proper differentiation.

Since in most tumors the functional inactivation of pRb is mediated by kinase activities amenable to pharmacological intervention, it seems reasonable to suggest that even a partial inhibition of the relevant kinases would rescue pRb functions. Our proposal that this would result in the recovery of differentiation programs in a large number of tumors adds a new rationale for testing cyclin-dependent kinase inhibitors as anti-cancer drugs and suggests new endpoints to evaluate their activities.

References

1. Hanahan D, Weinberg RA. The hallmarks of cancer. Cell 2000; 100(1):57-70.
2. Seremetis S, Inghirami G, Ferrero D et al. Transformation and plasmacytoid differentiation of EBV-infected human B lymphoblasts by ras oncogenes. Science 1989; 243(4891):660-663.
3. Cotran R, Kumar V, Collins T et al. Pathologic basis of disease. Philadelphia: Saunders, 1999.
4. Bartek J, Bartkova J, Lukas J. The retinoblastoma protein pathway in cell cycle control and cancer. Exp Cell Res 1997; 237(1):1-6.
5. Grana X, Garriga J, Mayol X. Role of the retinoblastoma protein family, pRB, p107 and p130 in the negative control of cell growth. Oncogene 1998; 17(25):3365-3383.
6. Sellers WR, Novitch BG, Miyake S et al. Stable binding to E2F is not required for the retinoblastoma protein to activate transcription, promote differentiation, and suppress tumor cell growth. Genes Dev 1998; 12(1):95-106.
7. Lipinski MM, Jacks T. The retinoblastoma gene family in differentiation and development. Oncogene 1999; 18(55):7873-7882.
8. Clarke AR, Maandag ER, van Roon M et al. Requirement for a functional Rb-1 gene in murine development. Nature 1992; 359(6393):328-330.
9. Jacks T, Fazeli A, Schmitt EM et al. Effects of an Rb mutation in the mouse. Nature 1992; 359(6393):295-300.
10. Lee EY, Chang CY, Hu N et al. Mice deficient for Rb are nonviable and show defects in neurogenesis and haematopoiesis. Nature 1992; 359(6393):288-294.
11. Morgenbesser SD, Williams BO, Jacks T et al. p53-dependent apoptosis produced by Rb-deficiency in the developing mouse lens. Nature 1994; 371(6492):72-74.
12. Schneider JW, Gu W, Zhu L et al. Reversal of terminal differentiation mediated by p107 in $Rb^{-/-}$ muscle cells. Science 1994; 264:1467-1471.
13. Zacksenhaus E, Jiang Z, Chung D et al. pRb controls proliferation, differentiation, and death of skeletal muscle cells and other lineages during embryogenesis. Genes Dev 1996; 10(23):3051-3064.
14. Chen PL, Riley DJ, Chen Y et al. Retinoblastoma protein positively regulates terminal adipocyte differentiation through direct interaction with C/EBPs. Genes Dev 1996; 10(21):2794-2804.
15. Classon M, Kennedy BK, Mulloy R et al. Opposing roles of pRB and p107 in adipocyte differentiation. Proc Natl Acad Sci USA 2000; 97(20):10826-10831.
16. Chen PL, Riley DJ, Chen-Kiang S et al. Retinoblastoma protein directly interacts with and activates the transcription factor NF-IL6. Proc Natl Acad Sci USA 1996; 93(1):465-469.
17. Bergh G, Ehinger M, Olsson I et al. Involvement of the retinoblastoma protein in monocytic and neutrophilic lineage commitment of human bone marrow progenitor cells. Blood 1999; 94(6):1971-1978.

18. Ji Y, Studzinski GP. Retinoblastoma protein and CCAAT/enhancer-binding protein beta are required for 1,25-dihydroxyvitamin D3-induced monocytic differentiation of HL60 cells. Cancer Res 2004; 64(1):370-377.
19. Gery S, Gombart AF, Fung YK et al. C/EBPε interacts with retinoblastoma and E2F1 during granulopoiesis. Blood 2004; 103(3):828-835.
20. Thomas DM, Carty SA, Piscopo DM et al. The retinoblastoma protein acts as a transcriptional coactivator required for osteogenic differentiation. Mol Cell 2001; 8(2):303-316.
21. Fletcher O, Easton D, Anderson K et al. Lifetime risks of common cancers among retinoblastoma survivors. J Natl Cancer Inst 2004; 96(5):357-363.
22. Williams BO, Schmitt EM, Remington L et al. Extensive contribution of Rb-deficient cells to adult chimeric mice with limited histopathological consequences. Embo J 1994; 13(18):4251-4259.
23. Maandag EC, van der Valk M, Vlaar M et al. Developmental rescue of an embryonic-lethal mutation in the retinoblastoma gene in chimeric mice. Embo J 1994; 13(18):4260-4268.
24. MacPherson D, Sage J, Crowley D et al. Conditional mutation of Rb causes cell cycle defects without apoptosis in the central nervous system. Mol Cell Biol 2003; 23(3):1044-1053.
25. Wu L, de Bruin A, Saavedra HI et al. Extra-embryonic function of Rb is essential for embryonic development and viability. Nature 2003; 421(6926):942-947.
26. Robinson GW, Wagner KU, Hennighausen L. Functional mammary gland development and oncogene-induced tumor formation are not affected by the absence of the retinoblastoma gene. Oncogene 2001; 20(48):7115-7119.
27. Day KC, McCabe MT, Zhao X et al. Rescue of embryonic epithelium reveals that the homozygous deletion of the retinoblastoma gene confers growth factor independence and immortality but does not influence epithelial differentiation or tissue morphogenesis. J Biol Chem 2002; 277(46):44475-84.
28. Cobrinik D, Lee MH, Hannon G et al. Shared role of the pRB-related p130 and p107 proteins in limb development. Genes Dev 1996; 10(13):1633-1644.
29. LeCouter JE, Kablar B, Hardy WR et al. Strain-dependent myeloid hyperplasia, growth deficiency, and accelerated cell cycle in mice lacking the Rb-related p107 gene. Mol Cell Biol 1998; 18(12):7455-7465.
30. LeCouter JE, Kablar B, Whyte PF et al. Strain-dependent embryonic lethality in mice lacking the retinoblastoma- related p130 gene. Development 1998; 125(23):4669-4679.
31. Novitch BG, Mulligan GJ, Jacks T et al. Skeletal muscle cells lacking the retinoblastoma protein display defects in muscle gene expression and accumulate in S and G2 phases of the cell cycle. J Cell Biol 1996; 135(2):441-456.
32. Condorelli GL, Testa U, Valtieri M et al. Modulation of retinoblastoma gene in normal adult hematopoiesis: Peak expression and functional role in advanced erythroid differentiation. Proc Natl Acad Sci USA 1995; 92(11):4808-4812.
33. De Bruin A, Wu L, Saavedra HI et al. Rb function in extraembryonic lineages suppresses apoptosis in the CNS of Rb-deficient mice. Proc Natl Acad Sci USA 2003; 100(11):6546-6551.
34. Rossi F, MacLean HE, Yuan W et al. p107 and p130 Coordinately regulate proliferation, Cbfa1 expression, and hypertrophic differentiation during endochondral bone development. Dev Biol 2002; 247(2):271-285.
35. Novitch BG, Spicer DB, Kim PS et al. pRb is required for MEF2-dependent gene expression as well as cell-cycle arrest during skeletal muscle differentiation. Curr Biol 1999; 9(9):449-459.
36. Puri PL, Iezzi S, Stiegler P et al. Class I histone deacetylases sequentially interact with MyoD and pRb during skeletal myogenesis. Mol Cell 2001; 8(4):885-897.
37. Miyake S, Sellers WR, Safran M et al. Cells degrade a novel inhibitor of differentiation with E1A-like properties upon exiting the cell cycle. Mol Cell Biol 2000; 20(23):8889-8902.
38. Morris EJ, Dyson NJ. Retinoblastoma protein partners. Adv Cancer Res 2001; 82:1-54.
39. Slack RS, El-Bizri H, Wong J et al. A critical temporal requirement for the retinoblastoma protein family during neuronal determination. J Cell Biol 1998; 140(6):1497-1509.
40. Marino S, Hoogervoorst D, Brandner S et al. Rb and p107 are required for normal cerebellar development and granule cell survival but not for Purkinje cell persistence. Development 2003; 130(15):3359-3368.
41. McCaffrey J, Yamasaki L, Dyson NJ et al. Disruption of retinoblastoma protein family function by human papillomavirus type 16 E7 oncoprotein inhibits lens development in part through E2F-1. Mol Cell Biol 1999; 19(9):6458-6468.
42. Gu W, Schneider JW, Condorelli G et al. Interaction of myogenic factors and the retinoblastoma protein mediates muscle cell commitment and differentiation. Cell 1993; 72:309-324.
43. Li FQ, Coonrod A, Horwitz M. Selection of a dominant negative retinoblastoma protein (RB) inhibiting satellite myoblast differentiation implies an indirect interaction between MyoD and RB. Mol Cell Biol 2000; 20(14):5129-5139.

44. Balsitis SJ, Sage J, Duensing S et al. Recapitulation of the effects of the human papillomavirus type 16 E7 oncogene on mouse epithelium by somatic Rb deletion and detection of pRb-independent effects of E7 in vivo. Mol Cell Biol 2003; 23(24):9094-9103.
45. Zhang J, Gray J, Wu L et al. Rb regulates proliferation and rod photoreceptor development in the mouse retina. Nat Genet 2004; 36(4):351-360.
46. Puigserver P, Ribot J, Serra F et al. Involvement of the retinoblastoma protein in brown and white adipocyte cell differentiation: Functional and physical association with the adipogenic transcription factor C/EBPalpha. Eur J Cell Biol 1998; 77(2):117-123.

CHAPTER 3

Regulation of DNA Replication by the Retinoblastoma Tumor Suppressor Protein

Erik S. Knudsen* **and Steven P. Angus**

The retinoblastoma gene product (RB) plays a critical role in the inhibition of cancer. This prototypical tumor suppressor was identified based on bi-alleleic inactiviation in the pediatric tumor retinoblastoma. Subsequently, it has become clear that multiple pathways lead to the functional inactivation of RB in most cancers. As such, substantial energy has been directed at delineating the mechanisms through which RB functions to limit tumorigenic potential. This review will focus on one facet of RB signaling—DNA replication control.

RB-Mediated Cell Cycle Control and Cancer

The functional inactivation of RB is an incredibly common feature of tumorigenesis.[1-7] Initially defined as the rate-limiting event in the development of retinoblastoma, loss of heterozygosity or genetic loss of *Rb* is observed in a number of cancers at relatively high frequency.[8,9] However, these malignancies only encompass one subset of those in which the function of RB is compromised. Viral oncoproteins of DNA tumor-viruses inactivate RB via direct binding in specific malignancies (e.g., the human papillomavirus E7 protein in cervical cancer).[10-12] Alternatively, RB can be inactivated via deregulated phosphorylation that is initiated via CDK4/cyclin D complexes.[5-7] These complexes harbor oncogenic activity and are kept in check by the tumor suppressor p16ink4a.[13] The presumption that these activities all function through RB is supported by a number of genetic and biochemical studies.[14,15]

RB and Cell Cycle Control

Our understanding of RB-dependent cell cycle control has gone through a number of refinements in recent years, as increasing evidence suggests a prominent role for RB in S-phase control.

In cells that have exited the cell cycle, RB is hypophosphorylated and functions to restrain cell cycle entry.[6,7,16] It is clear that RB must interact with specific binding proteins and assemble complexes to facilitate cell cycle inhibition.[17] Upon the engagement of mitogenic signaling, CDK4 or CDK6 complexes become active and initiate the phosphorylation of RB.[16] Subsequent phosphorylation via CDK2 complexes serve to maintain RB in its hyperphosphorylated state until mitosis, when RB is dephosphorylated through the action of a specific phosphatase.[16,18] The consequence of RB phosphorylation is the disruption of virtually all of its protein associations. As such, phosphorylation enables progression through subsequent phases of the cell cycle. This basic model fits essentially all data in the literature pertaining to cell cycle control by RB (Fig. 1). Critically, tumor alleles of RB from hereditary retinoblastoma or sporadic tumors are compromised for interacting with effector proteins or

*Corresponding Author: Erik S. Knudsen—Department of Cell Biology, University of Cincinnati College of Medicine, Vontz Center for Molecular Studies, Cincinnati, Ohio 45267, U.S.A. Email: erik.knudsen@uc.edu

Rb and Tumorigenesis, edited by Maurizio Fanciulli.

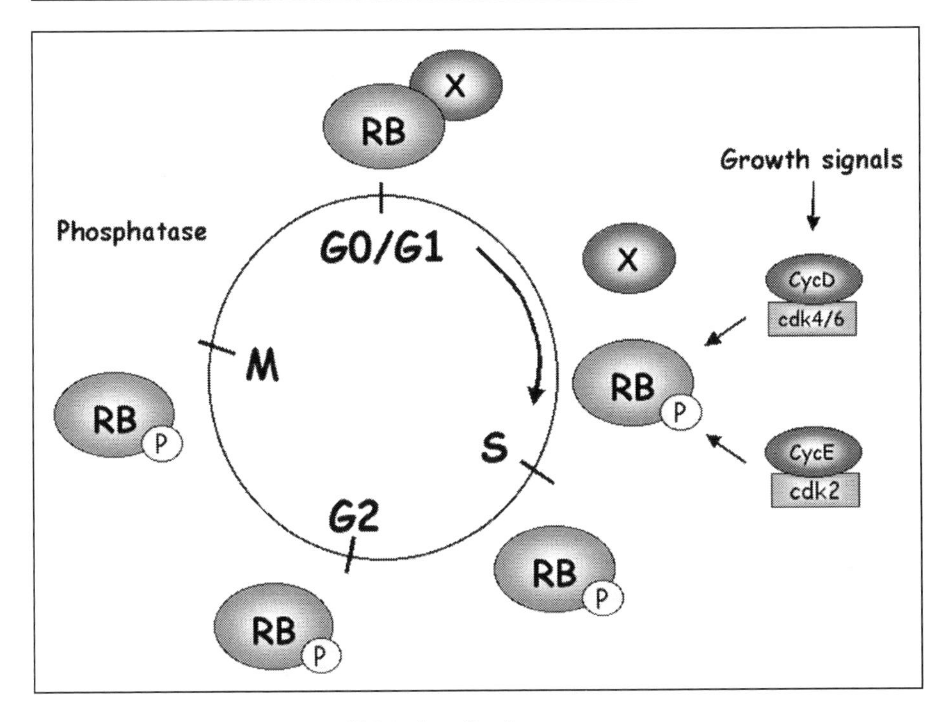

Figure 1. Schematic representation of RB in the cell cycle.

mediating cell cycle arrest.[6,7] In murine models of tumorigenesis, these alleles are similarly compromised for tumor suppression.[19-21] Correspondingly, the deregulated hyperphosphorylation of RB overcomes cell cycle inhibition in cell culture models and is a common feature in tumors arising through multiple mechanisms (i.e., amplification of cyclin D1 or loss of p16ink4a).[1,3,5-7,22]

G1/S Control by RB

Initial studies suggested that RB functions to constrain passage from G1 to S-phase. This model was supported by a series of studies performed in RB-deficient tumor cell lines, wherein the ectopic expression of RB resulted in the accumulation of cells with a 2N DNA content and the inhibition of replication.[23-25] This concept was reinforced by the findings that the presence of functional RB was important/required for the action of anti-mitogenic agents that elicit a G_1-arrest (e.g., TGF-beta in epithelial cells).[26,27] Lastly, it was found that loss of RB shifted the position of the restriction point in G_1, enabling cells to cross the commitment to S-phase earlier than cells proficient in RB.[28,29] For these reasons and the timing of RB hyperphosphorylation in mid/late G_1 it was postulated that RB represents a critical factor involved in integrating mitogenic signaling pathways in G_1 to enable progression into S-phase.

S-Phase Control by RB

That RB was solely a regulator of G_1 progression was challenged when it was found that the expression of constitutively active alleles of RB, that could not be inactivated by CDK/cyclin complexes, had the effect of arresting cells with both G1 and S-phase DNA contents.[30-32] Subsequently, it was found that agents that function in S-phase to inhibit DNA replication acted in an RB-dependent manner. For example, it has been found that the Rep78 protein of adeno-associated virus (AAV) inhibits DNA replication through the activation of endogenous RB in S-phase.[33] Additionally, DNA damaging agents elicit an RB dependent inhibition of

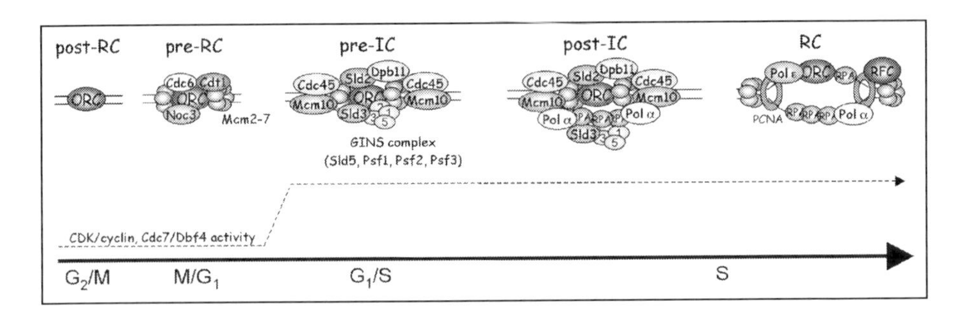

Figure 2. Schematic model for DNA replication.

S-phase progression.[34-37] This effect of RB is critical for the maintenance of ploidy as cells deficient in RB undergo DNA rereplication following DNA damage.[36,37] In parallel with these studies, it was found that compromising the function of the *Drosophila* RB related protein, RBf, resulted in aberrations in DNA replication and control of ploidy during development.[38] Such results indicate that S-phase deregulation, as occurs through RB loss, could contribute to genetic instability. In total these findings describe a critical role for RB in the control of S-phase.

DNA Replication: An Overview

Duplication of DNA during S-phase must occur once per cell cycle to ensure the propagation of genetic material during cellular division. Additionally, replication must be constrained to one complete doubling per cell cycle to maintain the appropriate genetic complement from one cellular generation to the next. Here, we will provide a brief summary of the process of DNA replication (Fig. 2). For a more in-depth treatise on DNA replication, there are several excellent reviews.[39-45]

PreRC

The initiation of DNA replication in eukaryotic cells occurs at multiple sites, called origins of replication (Fig. 2). While these sites in the budding yeast *Saccharomyces cerevisiae* are characterized by a short consensus sequence, origins in multicellular organisms are less well defined[46,47] and appear to be governed by a variety of factors in addition to primary sequence.[48,49] For example, the sites of DNA replication initiation in mammalian cells can be shifted by transcriptional state/chromatin structure or the availability of dNTPs.[50-52] The mechanisms dictating origin selection and usage are the subject of ongoing investigation in mammalian cells. Origins in all eukaryotes, however, serve as binding sites for the conserved hexameric origin recognition complex (ORC1-6), originally identified in *S. cerevisiae.*[53] ORC may be subject to cell cycle control, as mammalian Orc1 has been shown to dissociate and undergo proteolytic degradation after the initiation of DNA replication.[54-56] During exit from mitosis and progression through subsequent G_1-phase, prereplication complex (preRC) assembly occurs at prospective origins.[57,58] Formation of the preRC at origins has been termed "replication licensing" as it effectively enables initiation of DNA synthesis at a given site.[40] Initially, the Cdc6 and Cdt1 proteins bind to ORC and recruit the hexameric minichromosome maintenance (Mcm2-7) complex.[59-64] Recently, Zhang et al have identified an additional protein, Noc3p, required for preRC assembly in *S. pombe.*[65] The assembly of the preRC occurs in a CDK-independent manner, and begins as cells exit mitosis and enter G_1 of the subsequent cell cycle.

Initiation

While the key substrates have yet to be clearly identified, initiation of replication (also termed "origin firing") requires the activity of Cdc7 (also called DDK, or Dbf4-dependent kinase) and CDK complexes.[39] Several factors have been identified as substrates of CDK and

Cdc7 activity in vitro, but the consequence of these phosphorylation events on DNA replication has been equivocal.[66,67] For example, cyclin A-associated kinase activity can phosphorylate Cdc6, however, the consequence of this phosphorylation has been hotly debated without a clear consensus as to the significance of this event.[68-72] The recruitment of additional factors to a licensed origin establishes a preinitiation complex (preIC).[39,41,44] This stage of the DNA replication process has become increasingly complicated as multiple components have been recently described. Maturation of the preRC involves the binding of Mcm10, Cdc45 and Sld3, Sld2, Dpb11, and the GINS complex ("GINS" is a Japanese acronym for five, one, two, three signifying the complex of Sld5, Psf1, Psf2, and Psf3).[73-84] Ultimately, these factors give rise to the melting of DNA at the functional origin and provide the substrate for the initiation of DNA replication.

Replication Complexes

Progression of the bidirectional DNA replication forks requires numerous additional factors. Studies in yeast and *Xenopus* egg extract have suggested that ORC and Cdc6 are dispensable after replication initiation. In contrast, the Mcm complexes are required for ongoing DNA replication.[85,86] Several lines of evidence implicate Mcm complex as the replicative helicase.[86] The eukaryotic single-stranded binding protein, RPA, stabilizes the unwound double helical DNA and thus exposes the template strands. DNA polymerase alpha/primase (pol α/primase) complex initiates de novo DNA synthesis—first synthesizing a short RNA primer, which is extended by the DNA polymerase activity of the protein.[45] DNA pol α/primase is not capable of processive activity. Thus, the primer/template junction of the newly synthesized strand is recognized by the replication factor C (RFC) complex.[87] RFC loads the sliding clamp, proliferating cell nuclear antigen (PCNA), which topologically encircles the DNA. PCNA itself interacts with the processive DNA polymerases, δ and ε.[45] Efficient replication elongation may then ensue at the leading and lagging strands. Processing of the Okazaki fragments on the lagging strand occurs via PCNA-mediated recruitment of RNase HI and FEN1, which remove the RNA primer, and DNA ligase I, which joins the DNA fragments.[45] Together, these complexes mediate the complete replication of the genome and the subsequent processing of replication intermediates.

dNTP Pools

In addition to the factors required at the replication fork, DNA replication requires a supply of deoxyribonucleotides (dNTPs). The dNTP pool is regulated and maintained by the activity of several enzymes whose expression is cell cycle regulated (e.g., ribonucleotide reductase and thymidylate synthase).[88,89] Thus the pool of substrates for DNA replication is stimulated concomitant with entry into S-phase. Conversely, inhibition of dNTP pools represents one of the key means to block DNA replication (e.g., hydroxyurea). In mammalian cells, such blocks have been shown to stall DNA replication and induce alternate origin usage.[51] However, a recent study from Merrick et al found that dNTP depletion inhibited the firing of additional origins.[90] Critically, dNTP biology is one of the key targets for antimetabolite therapeutics (e.g., 5-fluorouracil and methotrexate) that have similarly profound effects on DNA replication.[91,92]

Given the highly orchestrated nature of DNA replication (Fig. 2), the inhibition of any one process in the pathway of replication will lead to a blockade of synthesis. Therefore, it is natural to speculate that inhibition of one or several of these replicative processes by RB is responsible for the observed inhibition of replication.

Direct Influence of RB on the DNA Replication Machinery

Currently, greater than 100 RB-interacting proteins have been identified by yeast two-hybrid assays, affinity column purification, and interaction screening of cDNA expression libraries.[17] RB has been shown to bind to several distinct DNA replication factors, suggesting that RB could directly inhibit the replication machinery. Additionally, RB has been found in sub-nuclear structures associated with replication and shown to directly associate with replication origins. Together, these studies make a good case for a direct influence of RB on the DNA replication machinery.

RB and Replication Foci

The RB protein contains a bipartite nuclear localization sequence and is found in the nucleus throughout the cell cycle.[93] However, a quite varied picture of RB has been observed when investigating the sub-nuclear localization of RB and its relationship with DNA replication complexes.

As discussed above, RB is phosphorylated as cells progress through G_1 to S-phase and early studies showed that this phosphorylation does not alter the localization of RB.[94] However, RB phosphorylation was associated with decreased retention of RB in a nuclear tethered form.[95] Specifically, hypophosphorylated RB is present in nuclei that have been extracted with a low concentration of nonionic detergent, whereas hyperphosphorylated RB is dissociated. The exact nature of this tetherered pool of RB and associated factors is not clear. Work from Mancini et al and others demonstrated that RB associated with the nuclear matrix and could bind lamin A/C in vitro,[96-98] suggesting that this interaction underlies the nuclear retention of RB. Conversely using a live-cell imaging approach it has been suggested that the retained form of RB is dependent on interactions with the E2F-family of transcription factors.[99] However, in all of these studies the association of RB with nuclear structures is attenuated as cells progress into S-phase.

Such findings would seemingly argue against the recruitment of RB to replication factories in S-phase, a time when RB is normally hyperphosphorylated. However, RB has been observed to specifically colocalize with sites of ongoing DNA replication. Kennedy et al found that RB colocalized with early-replicating regions, but not at later S-phase stages.[100] These perinucleolar foci were detectable during G_1 of the cell cycle and persisted into S-phase. Furthermore, it was suggested that RB function is required for spatial patterning of DNA replication. It should be noted, however, that this issue is not without controversy. A conflicting report was published in which RB did not colocalize with active replication foci, nuclear lamins, or replication proteins.[101] It is clear that further studies will be required to resolve these discrepancies.

RB Interactions with Origins

While the presence of RB as part of the replication machinery is somewhat clouded, there are studies which clearly support a role of RB at replication origins. Development of *Drosophila* involves rapid amplification of the chorion gene loci. Chorion genes, essential for egg shell formation, are present in a cluster and the origins of DNA replication have been identified.[102] Bosco et al showed that the *Drosophila* RB homologue (Rbf) immunoprecipitated with Orc subunits and Rbf bound near the chorion amplification origin with *Drosophila* E2F.[38] Importantly, an *Rbf* mutant displayed deregulated chorion gene amplification and genomic DNA replication. The use of E2F alleles and pharmacological inhibition of transcription supported the model that RB was functioning in this context through origin interactions and not through a transcriptional pathway.[38]

Investigation of the potential influence of RB on origin function is somewhat limited in mammalian cells due to the ambiguous nature of origins. In general, mammalian origins lack strong sequence specificity and are often characterized by wide initiation zones and varied origin utilization.[48] However, several mammalian origins have been identified and validated in multiple cellular systems.[49,103] Using chromatin immunoprecipitation, work from Avni et al convincingly demonstrate that upon the dephosphorylation/activation of endogenous RB with ionizing radiation, RB is recruited to chromatin in the proximity of origins.[37] This work provides a strong rationale for considering that RB could function in a direct manner to prevent origin firing and provides the impetus for defining molecular targets for RB directly involved in DNA replication at origins.

RB Interactions with Replication Factors

As noted above, RB has been shown to interact with many cellular proteins.[17] Amongst this group of proteins are several with critical roles in DNA replication that could represent one means through which RB could function to limit DNA replication.

MCM7

The Mcm7 subunit was found to interact with the N-terminus of RB in a yeast two-hybrid screen.[104] The functional significance of this interaction was demonstrated by the finding that the purified N-terminus of RB could strongly inhibit DNA replication in vitro, independently of transcription. The addition of excess purified Mcm7 could partially overcome the inhibitory effect of RB, supporting direct binding as the means of replication inhibition. As a component of the Mcm complex required for replication licensing, Mcm7 could therefore be subject to negative regulation by RB during G_1. In support of this hypothesis, Gladden and Diehl reported that CDK4/cyclin D1 complexes associate with Mcm7.[105] The catalytic activity of CDK4/cyclin D1 triggered the dissociation of Mcm7 from RB and perturbed the association of RB with chromatin. These studies support the notion that RB may act during G_1 to govern preRC formation and influence origin firing. While these data make a compelling argument for Mcm7 as a candidate target for RB-mediated inhibition of DNA replication, such an interaction is apparently not required for the inhibition of DNA replication by RB. Alleles of RB which lack the N-terminal Mcm7 binding site are capable of efficiently inhibiting DNA replication.[31,106] Furthermore, the N-terminal Mcm7-binding region of RB is not required for tumor suppression in mouse models.[21]

DNA Polymerase α

The DNA polymerase (pol) α protein was shown to interact with phosphorylated RB in pull down reactions.[107] Thus, DNA pol α. represents one of the few proteins reported to bind the phosphorylated form of RB. Intriguingly, hyperphosphorylated RB stimulated the activity of DNA pol α in in vitro replication reactions. The biological significance of this observation is unknown. Furthermore, the loss of RB has not been found to negatively influence DNA replication rates, as would be expected from these studies.

DNA Polymerase δ

In contrast with DNA pol α, the catalytic subunit of the processive DNA polymerase δ associates with hypophosphorylated RB.[108] However, RB also stimulated the in vitro activity of DNA pol δ. The significance of these interactions during S-phase has not been elucidated and would seemingly be at odds with those studies showing that the active/dephosphorylated form of RB is capable of inhibiting DNA replication and that the majority of RB protein is hyperphosphorylated in S-phase.

Replication Factor C

Lastly, the large subunit of replication factor C (RFC) was shown to interact with RB to promote cell survival following DNA damage.[109] Whether this interaction is relevant for normal S-phase progression or RB-dependent replication arrest has not been addressed. Since RFC has been shown to augment RB-mediated transcriptional repression,[110] its influence on replication could be through indirect transcriptional influences.

Collectively, these studies suggest that RB may exert a direct influence on DNA replication, either through preRC assembly, origin function, or the activity of replication fork enzymes. An additional, as of yet untested possibility, is that the recruitment of RB in the proximity of origins could serve to constrain chromatin structure through the recruitment of chromatin modifying factors that also function as transcriptional corepressors. In this context the same mechanisms utilized in transcriptional control would be employed to limit DNA replication through chromatin modifications in the proximity of replication origins.

RB-Mediated Transcriptional Repression and DNA Replication

A critical cellular target for RB is the E2F-family of transcription factors.[111] These factors were initially characterized through their ability to stimulate transcription of the adenoviral E2 promoter.[112] Subsequently, the first E2F gene was cloned based on its interaction with RB.[113-115] Today, the E2F family of transcription factors consists of seven members, which regulate the expression of a variety of genes.[116,117] Initially, many E2F target genes were identified based on promoter sequence or known G1/S activation.[118-120] However, with the advent of gene expression profiling by microarray analysis approximately 200 cellular genes are reproducibly regulated through the RB/E2F pathway.[121-124] A substantial number of E2F target genes are associated with DNA replication activities (Fig. 3). As described above, the phosphorylation of RB regulates its ability to bind virtually all of its cellular binding partners.[16] Hypophosphorylated RB is capable of associating with E2F family members.[111] Under normal proliferative conditions, phosphorylation of RB during G_1 results in the disruption of RB/E2F repressor complexes and permits the E2F-dependent activation of essential cell cycle genes.[111,125] The association of RB not only blocks the transactivation function of E2F, but also elicits active transcriptional repression of E2F-regulated promoters.[126-128] The repression of E2F-dependent promoters by RB depends on the simultaneous recruitment of corepressors through discrete interaction domains of RB. These corepressor molecules include histone deactylases (HDACs), components of the SWI/SNF chromatin remodeling complex, polycomb group proteins, and histone methylases.[129-136] As such, the activation of RB leads to the repression of E2F target genes that are required for replication (Fig. 3).[111,137] Importantly, this influence of RB on transcription is not restricted to specific phases of the cell cycle and RB-mediated repression is effective in S-phase of the cell cycle.[138] E2F-dependent transcriptional repression is required for cell cycle inhibition, as multiple groups have shown that the specific disruption of E2F-repressor complexes is sufficient to bypass RB-mediated arrest.[27,133,134,139] These results suggest that either transcriptional control is responsible for replication arrests or that E2F is an intimate mediator of RB direct action at origins. Consistent with a transcriptional role for RB in replication control, the disruption of transcriptionally activating E2F genes in mice results in a blockade in S-phase progression which is associated with decreased E2F-target gene expression.[140,141] Similarly, in *Drosophila* the aberrant replication occurring through the loss of Rbf, is suppressed by lowering the gene dosage of E2F stimulated genes.[142] Thus, RB-mediated S-phase inhibition may be achieved via a transcriptional pathway, wherein transcriptional repression of key targets achieves the cessation of DNA replication.

Targets of RB-Mediated Transcriptional Repression

Genes regulated by RB/E2F that are involved in DNA replication (Fig. 3) can be divided into several categories: (1) Cdc7/Dbf4 kinase complexes, (2) structural and enzymatic components of the DNA replication machinery, (3) CDK/cyclins and (4) dNTP synthetic enzymes.

Cdc7/Dbf4 Transcriptional Regulation

Initiation of DNA replication requires the activity of the Cdc7/Dbf4 complex.[39] Levels of Cdc7 do not change remarkably during the cell cycle.[143] In contrast, Dbf4 expression is responsive to E2F and oscillates during the cell cycle.[144] The Cdc7/Dbf4 complex likely cooperates with CDK activity, phosphorylating Mcm subunits to trigger the loading of Cdc45.[66,145,146] During RB-mediated cell cycle arrest, the protein levels of Dbf4 remain unchanged.[147] However, the effect of RB on Cdc7/Dbf4 complex formation and activity is unknown. Constanzo et al have recently identified a DNA damage checkpoint, dependent on Cdc7, that acts to inhibit Cdc45 loading.[148] Thus, both CDK and Cdc7 activities are targeted by DNA damage checkpoints to regulate DNA replication. A Dbf4-related factor (Drf1) was identified as an additional activator of Cdc7.[149] Although Drf1 is not apparently required for initiation of replication, it associates with chromatin in response to aphidicolin-induced replication blocks.[150] The effect of RB arrest on Drf1 has not been explored.

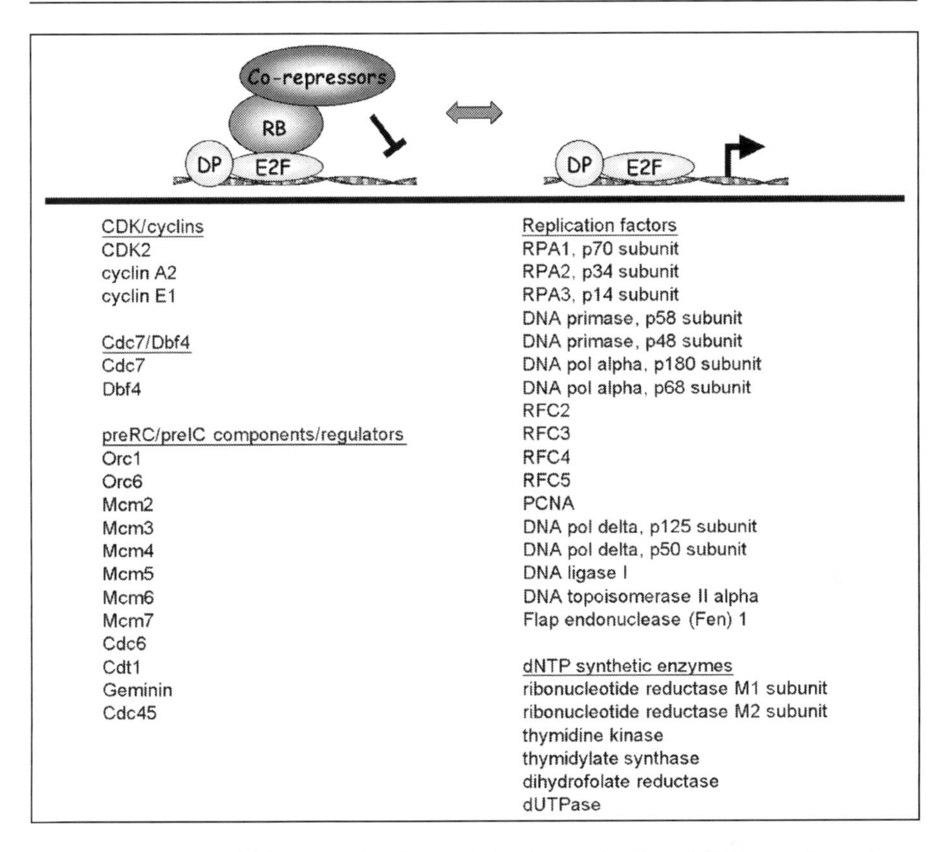

Figure 3. Schematic model for RB-mediated transcriptional repression through E2F transcription factors and identified targets of RB/E2F transcriptional control.

CDK/Cyclin Subunits

Multiple studies have demonstrated regulation of CDK2, cyclin E, and cyclin A through the RB/E2F pathway. Specifically, the targeted disruption of RB leads to increased CDK2, cyclin E and cyclin A protein levels.[28,151,152] Conversely, the activation of RB has been shown to enforce the downregulation of these factors.[27,31,32,35,153] While the details vary between laboratories, one constant in this regulation is the rapid and dramatic attenuation of cyclin A expression following the activation of RB. In fact, the attenuation of cyclin A occurs with kinetics rapid enough to establish it as an intermediate in RB mediated S-phase inhibition. The reason for this rapid influence is a coordinate effect of RB on cyclin A promoter activity and protein stability.[154]

CDK2, cyclin E, and cyclin A have been strongly implicated in progression through S-phase.[67] The activity of CDK2 increases as cells progress into S-phase and replication is stimulated in vitro through the addition of purified CDK2 complexes,[155,156] while inhibition of CDK2, cyclin E, or cyclin A activity was associated with the inhibition of DNA replication.[157-161] Such analyses suggested that both the CDK2/cyclin E and CDK2/cyclin A complexes are required for progression through S-phase. In fact, Coverley et al recently provided in vitro evidence for separable roles of cyclin E and cyclin A in DNA replication.[162] However, the critical nature of CDK2-associated activity has been challenged by the generation of viable CDK2-null mice, suggesting that additional kinases compensate for CDK2 loss.[163] Similarly, several cancer cell lines have been shown to proliferate in the absence of CDK2.[164] These studies are in agreement with earlier work from DeGregori et al showing that E2F activity could stimulate S-phase entry despite CDK2 inhibition.[165] Recently, Geng et al reported that the targeted disruption of the cyclin E1 and E2 genes did not inhibit mouse development or prevent DNA replication.[166] However, cyclin E-deficient cells were unable to reenter the cell cycle from G_0 and failed to load Mcm complex onto chromatin, suggesting that cyclin E plays important roles in specific replication outcomes. In contrast with CDK2 and cyclin E wherein loss can be tolerated, cyclin A appears to be required and nonredundant for replication control.[106,161,167,168] Given the tight regulation of cyclin A levels by RB, this suggests at least one critical mechanism through which RB could function to inhibit replication.

How cyclin A influences the replication machinery is only partially determined. Cyclin A has been observed at replication foci and can phosphorylate replication fork proteins such as RPA and DNA polymerases α and δ.[169-171] However, the consequence of these phosphorylation events during cellular DNA replication has not been revealed. Following the analysis of numerous replication factors, a functional target for the RB-repressed cyclin A was identified.[106] S-phase inhibition mediated by RB resulted in the specific disruption of PCNA association with chromatin.[106] Under these same conditions, components of the preRC and preIC, were unaffected. In keeping with expectation, DNA ligase I, which is dependent on PCNA for activity, was also soluble during RB-mediated arrest.[147] These results delineated a pathway through which RB functions transcriptionally to impact the DNA replication machinery via cyclin A kinase activity. Critically, the specific substrate regulating the loading/maintenance of PCNA on chromatin has yet to be illuminated, as the influence of cyclin A is apparently not through the direct phosphorylation of PCNA.

Replication Factors

The majority of proteins required for preRC formation and at DNA replication forks are subject to E2F control (Fig. 3). Thus, the cell cycle arrest induced by active, hypophosphorylated RB could potentially depend on the transcriptional repression of any, or all, of these critical genes. Correspondingly, it has been demonstrated that aberrant replication occurring through loss of Rbf in *Drosophila*, can be attenuated via the lowering of MCM gene dosage. While RNA levels of many of these genes are repressed by RB,[138] analysis of protein levels has not revealed significant attenuation concomitant with the inhibition of DNA replication.[147] This finding is consistent with the observation that most replication factors have relatively long half-lives and thus while RNA levels are diminished there is not a major change in protein levels. The levels of Mcm proteins, Dbf4, PCNA, and other replication components are reduced with delayed kinetics (36-48 hours after active RB expression). Importantly, this phenomenon also occurs as part of the RB-dependent response to DNA damage, wherein the conditional inactivation of RB is sufficient to alleviate the attenuation of replication factor expression and allow chronically arrested cells to reenter the cell cycle.[147] The biological significance of these distinct effects of RB on the replication machinery has yet to be explored, but is hypothesized to produce a cell cycle state from which cell cycle entry is further inhibited.

dNTP Synthetic Enzymes

In addition to the protein components that drive DNA replication, the duplication of the genetic material requires available DNA monomers. The deoxyribonucleotide (dNTP) pool must be carefully balanced to ensure the fidelity of DNA replication.[88,89] Additionally, the dNTP pool undergoes changes coincident with cell cycle phase. This observation is consistent with the fact that E2F regulates the expression of several dNTP metabolic enzymes—ribonucleotide reductase (RNR), thymidylate synthase (TS), dihydrofolate reductase (DHFR), and thymidine kinase (TK).[118,172] Almasan et al initially observed that the levels of DHFR and TS were elevated in *Rb-/-* MEFs.[172] Furthermore, RB-null cells exhibited resistance to methotrexate, a DHFR-specific inhibitor, when compared to wild-type cells. TS and DHFR displayed a divergent response to active RB signaling, similar to the effect on cyclin E versus cyclin A.[173]

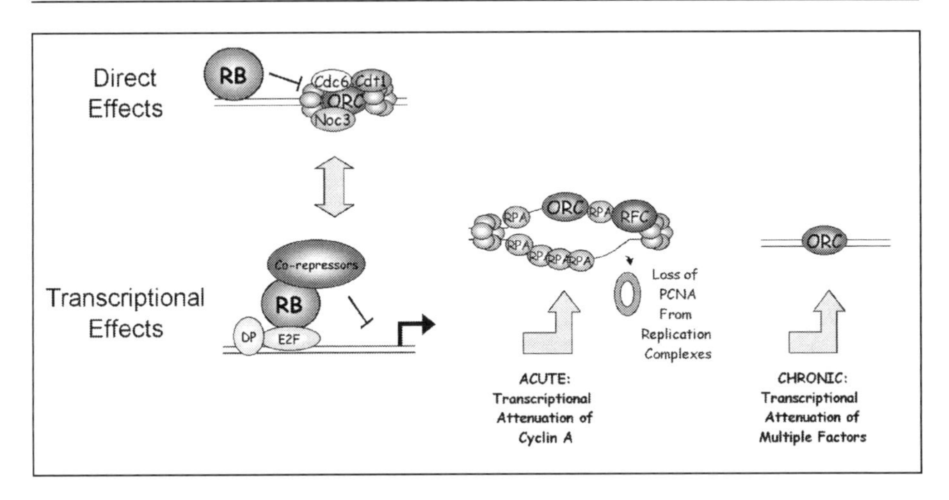

Figure 4. Model for combined direct and transcriptional effects of RB upon the replication machinery.

While both enzymes are E2F-regulated, only TS was strongly downregulated during RB-mediated cell cycle arrest. In addition to TS, the expression of RNR subunits R1 and R2 were substantially repressed. The loss of TS and RNR occurred rapidly (similar to cyclin A) and was correlated with a reduction in available dNTPs.[173] However, the continued presence of active RB led to the gradual attenuation of DHFR. The integration of S-phase control with dNTP pool regulation has become increasingly clear. For instance, the cell cycle regulation of RNR-R2 occurs at both the transcriptional and post-translational levels, subject to E2F control and proteolytic degradation by the anaphase-promoting complex.[174,175] Recently, a p53 inducible gene encoding an alternate R2 subunit of RNR (p53R2) was identified.[176,177] While the effect of RB on p53R2 has not been examined, it is likely that dNTP pool depletion by RB is not of sufficient magnitude to prevent DNA repair. Further studies are necessary to elucidate the function of RB in dNTP pool maintenance and S-phase progression, but the metabolic enzymes are clearly important targets of this tumor suppressor.

Together, the studies described make a case for transcriptional control being one means through which RB inhibits DNA replication. This response apparently, is bi-phasic in nature as determined by the relative stability of target proteins. RB initially acts through the attenuation of cyclin A and in a delayed manner leads to the loss of a large fraction of the replication machinery.

Multiple Targets and Model for RB-Mediated Replication Control

Why RB targets replication in multiple ways is an enigma. However, it is possible that RB enacts a rapid nontranscriptional response to inhibit origin firing. Such a response would be postulated to involve the association of RB with origins and a corresponding inhibition of replication initiation at origins. Coordinate transcriptional repression of cyclin A would stall replication while maintaining the replication machinery largely intact. Under such conditions, the inhibition of replication would be expected to be readily reversible. In fact, this is the case wherein cells acutely arrested by RB can be induced to reenter the cell cycle by removing RB. An important consideration for this type of model, is that many of the replication factors are actually bi-functional molecules that play roles in DNA damage repair. As such, it would be important to maintain these factors to facilitate repair with paused replication. In contrast, under conditions of persistent RB activation, possibly in response to more severe genotoxic stress, a replicative exit may be induced wherein the repertoire of replication factors is severely limited. Such a block would be expected to require considerably more factors to overcome and be correspondingly more difficult to bypass.

Considerations for the Future

While progress has been made in determining how RB functions in regulating DNA synthesis, key questions remain.

Impact of RB Loss on Replication

The loss of RB has significant effects on the levels of requisite replication factors. Specifically, numerous replication factors become deregulated. The extent to which these factors participate in replication (or stimulate replication rate) is completely unclear. Similarly, whether RB loss actually enhances/modifies origin usage is unknown.

Role of RB in Replication Fidelity

Several studies have shown that loss of RB can lead to changes in ploidy following genotoxic stress. These results indicate that RB is important for coupling the completion of mitosis with S-phase. However, whether RB is essential for maintaining the fidelity of replication and controlling the licensing of replication on unreplicated DNA has not been explicitly demonstrated.

Role of Replication Control as a Therapeutic Target

The relaxation of replication control in Rb-deficient cells is an important consideration for therapeutic strategies. A number of agents are more effective at killing cells or inducing secondary forms of DNA damage as cells progress through replication. Such agents would be expected to have enhanced activity against RB deficient cells through replication associated DNA damage.

Role of S-Phase Control in Tumor Suppression

While it is appealing to speculate on the importance of S-phase control as a critical means through which RB inhibits tumorigenesis, there is no data that has shown this to be the case. Clearly, the definition of the mechanisms through which RB impacts S-phase will provide the means to develop models to explicitly test the importance of these discrete pathways in tumorigenesis.

Role of RB in Intra-S-Phase DNA Damage Response

It is well known that DNA damage leads to a rapid inhibition of DNA synthesis (e.g., radio-resistant DNA synthesis). At present, the role of RB in this rapid, often transient response to DNA damage has not been determined. Since all the evidence suggests that RB must be dephosphorylated to inhibit S-phase, it suggests that rapid loss of phosphate from RB would be required for any rapid inhibition of DNA replication. Whether specific pools of RB molecules in the proximity of origins or DNA-lesions is dephosphorylated more rapidly has not been elucidated.

Summary

RB is a critical tumor suppressor targeted at high frequency in human cancers. As part of its mode of action RB participates in the regulation of DNA replication. In the absence of functional RB aberrant replication cycles occur and genotoxic stresses are compromised for inhibiting of DNA replication. Varied mechanisms through which RB inhibits S-phase have been described and provide evidence for temporally regulated stalling of replication initiation followed by a stable replicative exit.

Acknowledgements

In this review we tried to encompass as much of the literature as possible. To those whose fine work we may have overlooked we apologize. We received expert editorial commentary from Dr. Giovanni Bosco (University of Arizona), Dr. James DeGregori (University of Colorado Health Science Center), Dr. Cyrus Vaziri (Boston University), Dr. Jean Cook (University of North Carolina) and Dr. Karen Knudsen (University of Cincinnati). We are grateful to members of the Knudsens' and Nevins' laboratories for thought-provoking discussions and particularly Emily Bosco, Christopher Mayhew, David Solomon, and Wes Braden for review of the manuscript.

References

1. Bartkova J, Lukas J, Bartek J. Aberrations of the G1- and G1/S-regulating genes in human cancer. Prog Cell Cycle Res 1997; 3:211-220.
2. Classon M, Harlow E. The retinoblastoma tumour suppressor in development and cancer. Nat Rev Cancer 2002; 2(12):910-917.
3. Kaelin Jr WG. Alterations in G1/S cell-cycle control contributing to carcinogenesis. Ann NY Acad Sci 1997; 833:29-33.
4. Nevins JR. The Rb/E2F pathway and cancer. Hum Mol Genet 2001; 10(7):699-703.
5. Sherr CJ. Cancer cell cycles. Science 1996; 274(5293):1672-1677.
6. Wang JY, Knudsen ES, Welch PJ. The retinoblastoma tumor suppressor protein. Adv Cancer Res 1994; 64:25-85.
7. Weinberg RA. The retinoblastoma protein and cell cycle control. Cell 1995; 81(3):323-330.
8. Friend SH, Bernards R, Rogelj S et al. A human DNA segment with properties of the gene that predisposes to retinoblastoma and osteosarcoma. Nature 1986; 323(6089):643-646.
9. Lee WH, Bookstein R, Hong F et al. Human retinoblastoma susceptibility gene: Cloning, identification, and sequence. Science 1987; 235(4794):1394-1399.
10. DeCaprio JA, Ludlow JW, Figge J et al. SV40 large tumor antigen forms a specific complex with the product of the retinoblastoma susceptibility gene. Cell 1988; 54(2):275-283.
11. Whyte P, Buchkovich KJ, Horowitz JM et al. Association between an oncogene and an anti-oncogene: The adenovirus E1A proteins bind to the retinoblastoma gene product. Nature 1988; 334(6178):124-129.
12. Munger K, Werness BA, Dyson N et al. Complex formation of human papillomavirus E7 proteins with the retinoblastoma tumor suppressor gene product. Embo J 1989; 8(13):4099-4105.
13. Sherr CJ, Roberts JM. CDK inhibitors: Positive and negative regulators of G1-phase progression. Genes Dev 1999; 13(12):1501-1512.
14. Lukas J, Bartkova J, Rohde M et al. Cyclin D1 is dispensable for G1 control in retinoblastoma gene-deficient cells independently of cdk4 activity. Mol Cell Biol 1995; 15(5):2600-2611.
15. Lukas J, Parry D, Aagaard L et al. Retinoblastoma-protein-dependent cell-cycle inhibition by the tumour suppressor p16. Nature 1995; 375(6531):503-506.
16. Mittnacht S. Control of pRB phosphorylation. Curr Opin Genet Dev 1998; 8(1):21-27.
17. Morris EJ, Dyson NJ. Retinoblastoma protein partners. Adv Cancer Res 2001; 82:1-54.
18. Lundberg AS, Weinberg RA. Functional inactivation of the retinoblastoma protein requires sequential modification by at least two distinct cyclin-cdk complexes. Mol Cell Biol 1998; 18(2):753-761.
19. Vooijs M, Berns A. Developmental defects and tumor predisposition in Rb mutant mice. Oncogene 1999; 18(38):5293-5303.
20. Nikitin AY, Riley DJ, Lee WH. A paradigm for cancer treatment using the retinoblastoma gene in a mouse model. Ann NY Acad Sci 1999; 886:12-22.
21. Yang H, Williams BO, Hinds PW et al. Tumor suppression by a severely truncated species of retinoblastoma protein. Mol Cell Biol 2002; 22(9):3103-3110.
22. Sherr CJ, McCormick F. The RB and p53 pathways in cancer. Cancer Cell 2002; 2(2):103-112.
23. Huang HJ, Yee JK, Shew JY et al. Suppression of the neoplastic phenotype by replacement of the RB gene in human cancer cells. Science 1988; 242(4885):1563-1566.
24. Goodrich DW, Wang NP, Qian YW et al. The retinoblastoma gene product regulates progression through the G1 phase of the cell cycle. Cell 1991; 67(2):293-302.
25. Qin XQ, Chittenden T, Livingston DM et al. Identification of a growth suppression domain within the retinoblastoma gene product. Genes Dev 1992; 6(6):953-964.
26. Herrera RE, Makela TP, Weinberg RA. TGF beta-induced growth inhibition in primary fibroblasts requires the retinoblastoma protein. Mol Biol Cell 1996; 7(9):1335-1342.
27. Zhang HS, Postigo AA, Dean DC. Active transcriptional repression by the Rb-E2F complex mediates G1 arrest triggered by p16INK4a, TGFbeta, and contact inhibition. Cell 1999; 97(1):53-61.
28. Herrera RE, Sah VP, Williams BO et al. Altered cell cycle kinetics, gene expression, and G1 restriction point regulation in Rb-deficient fibroblasts. Mol Cell Biol 1996; 16(5):2402-2407.
29. Hurford Jr RK, Cobrinik D, Lee MH et al. pRB and p107/p130 are required for the regulated expression of different sets of E2F responsive genes. Genes Dev 1997; 11(11):1447-1463.
30. Chew YP, Ellis M, Wilkie S et al. pRB phosphorylation mutants reveal role of pRB in regulating S phase completion by a mechanism independent of E2F. Oncogene 1998; 17(17):2177-2186.

31. Knudsen ES, Buckmaster C, Chen TT et al. Inhibition of DNA synthesis by RB: Effects on G1/S transition and S-phase progression. Genes Dev 1998; 12(15):2278-2292.
32. Lukas J, Sorensen CS, Lukas C et al. p16INK4a, but not constitutively active pRb, can impose a sustained G1 arrest: Molecular mechanisms and implications for oncogenesis. Oncogene 1999; 18(27):3930-3935.
33. Saudan P, Vlach J, Beard P. Inhibition of S-phase progression by adeno-associated virus Rep78 protein is mediated by hypophosphorylated pRb. Embo J 2000; 19(16):4351-4361.
34. Harrington EA, Bruce JL, Harlow E et al. pRB plays an essential role in cell cycle arrest induced by DNA damage. Proc Natl Acad Sci USA 1998; 95(20):11945-11950.
35. Knudsen KE, Booth D, Naderi S et al. RB-dependent S-phase response to DNA damage. Mol Cell Biol 2000; 20(20):7751-7763.
36. Lan Z, Sever-Chroneos Z, Strobeck MW et al. DNA damage invokes mismatch repair-dependent cyclin D1 attenuation and retinoblastoma signaling pathways to inhibit CDK2. J Biol Chem 2002; 277(10):8372-8381.
37. Avni D, Yang H, Martelli F et al. Active localization of the retinoblastoma protein in chromatin and its response to S phase DNA damage. Mol Cell 2003; 12(3):735-746.
38. Bosco G, Du W, Orr-Weaver TL. DNA replication control through interaction of E2F-RB and the origin recognition complex. Nat Cell Biol 2001; 3(3):289-295.
39. Bell SP, Dutta A. DNA replication in eukaryotic cells. Annu Rev Biochem 2002; 71:333-374.
40. Blow JJ, Hodgson B. Replication licensing—defining the proliferative state? Trends Cell Biol 2002; 12(2):72-78.
41. Diffley JF, Labib K. The chromosome replication cycle. J Cell Sci 2002; 115(Pt 5):869-872.
42. Kelly TJ, Brown GW. Regulation of chromosome replication. Annu Rev Biochem 2000; 69:829-880.
43. Lei M, Tye BK. Initiating DNA synthesis: From recruiting to activating the MCM complex. J Cell Sci 2001; 114(Pt 8):1447-1454.
44. Mendez J, Stillman B. Perpetuating the double helix: Molecular machines at eukaryotic DNA replication origins. Bioessays 2003; 25(12):1158-1167.
45. Waga S, Stillman B. The DNA replication fork in eukaryotic cells. Annu Rev Biochem 1998; 67:721-751.
46. Vashee S, Cvetic C, Lu W et al. Sequence-independent DNA binding and replication initiation by the human origin recognition complex. Genes Dev 2003; 17(15):1894-1908.
47. Schaarschmidt D, Baltin J, Stehle IM et al. An episomal mammalian replicon: Sequence-independent binding of the origin recognition complex. Embo J 2004; 23(1):191-201.
48. Gilbert DM. Making sense of eukaryotic DNA replication origins. Science 2001; 294(5540):96-100.
49. Todorovic V, Falaschi A, Giacca M. Replication origins of mammalian chromosomes: The happy few. Front Biosci 1999; 4:D859-868.
50. Vogelauer M, Rubbi L, Lucas I et al. Histone acetylation regulates the time of replication origin firing. Mol Cell 2002; 10(5):1223-1233.
51. Anglana M, Apiou F, Bensimon A et al. Dynamics of DNA replication in mammalian somatic cells: Nucleotide pool modulates origin choice and interorigin spacing. Cell 2003; 114(3):385-394.
52. McNairn AJ, Gilbert DM. Epigenomic replication: Linking epigenetics to DNA replication. Bioessays 2003; 25(7):647-656.
53. Bell SP, Stillman B. ATP-dependent recognition of eukaryotic origins of DNA replication by a multiprotein complex. Nature 1992; 357(6374):128-134.
54. Kreitz S, Ritzi M, Baack M et al. The human origin recognition complex protein 1 dissociates from chromatin during S phase in HeLa cells. J Biol Chem 2001; 276(9):6337-6342.
55. Mendez J, Zou-Yang XH, Kim SY et al. Human origin recognition complex large subunit is degraded by ubiquitin-mediated proteolysis after initiation of DNA replication. Mol Cell 2002; 9(3):481-491.
56. DePamphilis ML. The 'ORC cycle': A novel pathway for regulating eukaryotic DNA replication. Gene 2003; 310:1-15.
57. Mendez J, Stillman B. Chromatin association of human origin recognition complex, cdc6, and minichromosome maintenance proteins during the cell cycle: Assembly of prereplication complexes in late mitosis. Mol Cell Biol 2000; 20(22):8602-8612.
58. Dimitrova DS, Prokhorova TA, Blow JJ et al. Mammalian nuclei become licensed for DNA replication during late telophase. J Cell Sci 2002; 115(Pt 1):51-59.
59. Cocker JH, Piatti S, Santocanale C et al. An essential role for the Cdc6 protein in forming the prereplicative complexes of budding yeast. Nature 1996; 379(6561):180-182.
60. Tanaka T, Knapp D, Nasmyth K. Loading of an Mcm protein onto DNA replication origins is regulated by Cdc6p and CDKs. Cell 1997; 90(4):649-660.
61. Liang C, Stillman B. Persistent initiation of DNA replication and chromatin-bound MCM proteins during the cell cycle in cdc6 mutants. Genes Dev 1997; 11(24):3375-3386.

62. Maiorano D, Moreau J, Mechali M. XCDT1 is required for the assembly of prereplicative complexes in Xenopus laevis. Nature 2000; 404(6778):622-625.
63. Nishitani H, Lygerou Z, Nishimoto T et al. The Cdt1 protein is required to license DNA for replication in fission yeast. Nature 2000; 404(6778):625-628.
64. Tanaka S, Diffley JF. Interdependent nuclear accumulation of budding yeast Cdt1 and Mcm2-7 during G1 phase. Nat Cell Biol 2002; 4(3):198-207.
65. Zhang Y, Yu Z, Fu X et al. Noc3p, a bHLH protein, plays an integral role in the initiation of DNA replication in budding yeast. Cell 2002; 109(7):849-860.
66. Masai H, Arai K. Cdc7 kinase complex: A key regulator in the initiation of DNA replication. J Cell Physiol 2002; 190(3):287-296.
67. Woo RA, Poon RY. Cyclin-dependent kinases and S phase control in mammalian cells. Cell Cycle 2003; 2(4):316-324.
68. Jiang W, Wells NJ, Hunter T. Multistep regulation of DNA replication by Cdk phosphorylation of HsCdc6. Proc Natl Acad Sci USA 1999; 96(11):6193-6198.
69. Petersen BO, Lukas J, Sorensen CS et al. Phosphorylation of mammalian CDC6 by cyclin A/CDK2 regulates its subcellular localization. Embo J 1999; 18(2):396-410.
70. Herbig U, Griffith JW, Fanning E. Mutation of cyclin/cdk phosphorylation sites in HsCdc6 disrupts a late step in initiation of DNA replication in human cells. Mol Biol Cell 2000; 11(12):4117-4130.
71. Cook JG, Park CH, Burke TW et al. Analysis of Cdc6 function in the assembly of mammalian prereplication complexes. Proc Natl Acad Sci USA 2002; 99(3):1347-1352.
72. Alexandrow MG, Hamlin JL. Cdc6 chromatin affinity is unaffected by serine-54 phosphorylation, S-phase progression, and overexpression of cyclin A. Mol Cell Biol 2004; 24(4):1614-1627.
73. Homesley L, Lei M, Kawasaki Y et al. Mcm10 and the MCM2-7 complex interact to initiate DNA synthesis and to release replication factors from origins. Genes Dev 2000; 14(8):913-926.
74. Wohlschlegel JA, Dhar SK, Prokhorova TA et al. Xenopus Mcm10 binds to origins of DNA replication after Mcm2-7 and stimulates origin binding of Cdc45. Mol Cell 2002; 9(2):233-240.
75. Zou L, Stillman B. Formation of a preinitiation complex by S-phase cyclin CDK-dependent loading of Cdc45p onto chromatin. Science 1998; 280(5363):593-596.
76. Tercero JA, Labib K, Diffley JF. DNA synthesis at individual replication forks requires the essential initiation factor Cdc45p. Embo J 2000; 19(9):2082-2093.
77. Kamimura Y, Tak YS, Sugino A et al. Sld3, which interacts with Cdc45 (Sld4), functions for chromosomal DNA replication in Saccharomyces cerevisiae. Embo J 2001; 20(8):2097-2107.
78. Nakajima R, Masukata H. SpSld3 is required for loading and maintenance of SpCdc45 on chromatin in DNA replication in fission yeast. Mol Biol Cell 2002; 13(5):1462-1472.
79. Wang H, Elledge SJ. DRC1, DNA replication and checkpoint protein 1, functions with DPB11 to control DNA replication and the S-phase checkpoint in Saccharomyces cerevisiae. Proc Natl Acad Sci USA 1999; 96(7):3824-3829.
80. Kamimura Y, Masumoto H, Sugino A et al. Sld2, which interacts with Dpb11 in Saccharomyces cerevisiae, is required for chromosomal DNA replication. Mol Cell Biol 1998; 18(10):6102-6109.
81. Araki H, Leem SH, Phongdara A et al. Dpb11, which interacts with DNA polymerase II(epsilon) in Saccharomyces cerevisiae, has a dual role in S-phase progression and at a cell cycle checkpoint. Proc Natl Acad Sci USA 1995; 92(25):11791-11795.
82. Masumoto H, Sugino A, Araki H. Dpb11 controls the association between DNA polymerases alpha and epsilon and the autonomously replicating sequence region of budding yeast. Mol Cell Biol 2000; 20(8):2809-2817.
83. Kubota Y, Takase Y, Komori Y et al. A novel ring-like complex of Xenopus proteins essential for the initiation of DNA replication. Genes Dev 2003; 17(9):1141-1152.
84. Takayama Y, Kamimura Y, Okawa M et al. GINS, a novel multiprotein complex required for chromosomal DNA replication in budding yeast. Genes Dev 2003; 17(9):1153-1165.
85. Tye BK. MCM proteins in DNA replication. Annu Rev Biochem 1999; 68:649-686.
86. Forsburg SL. Eukaryotic MCM proteins: Beyond replication initiation. Microbiol Mol Biol Rev 2004; 68(1):109-131, table of contents.
87. Mossi R, Hubscher U. Clamping down on clamps and clamp loaders—the eukaryotic replication factor C. Eur J Biochem 1998; 254(2):209-216.
88. Mathews CK, Ji J. DNA precursor asymmetries, replication fidelity, and variable genome evolution. Bioessays 1992; 14(5):295-301.
89. Reichard P. Interactions between deoxyribonucleotide and DNA synthesis. Annu Rev Biochem 1988; 57:349-374.
90. Merrick CJ, Jackson D, Diffley JF. Visualization of altered replication dynamics after DNA damage in human cells. J Biol Chem 2004; 279(19):20067-20075.

91. Longley DB, Harkin DP, Johnston PG. 5-fluorouracil: Mechanisms of action and clinical strategies. Nat Rev Cancer 2003; 3(5):330-338.
92. Schweitzer BI, Dicker AP, Bertino JR. Dihydrofolate reductase as a therapeutic target. Faseb J 1990; 4(8):2441-2452.
93. Zacksenhaus E, Bremner R, Phillips RA et al. A bipartite nuclear localization signal in the retinoblastoma gene product and its importance for biological activity. Mol Cell Biol 1993; 13(8):4588-4599.
94. Lee WH, Shew JY, Hong FD et al. The retinoblastoma susceptibility gene encodes a nuclear phosphoprotein associated with DNA binding activity. Nature 1987; 329(6140):642-645.
95. Mittnacht S, Weinberg RA. G1/S phosphorylation of the retinoblastoma protein is associated with an altered affinity for the nuclear compartment. Cell 1991; 65(3):381-393.
96. Mancini MA, Shan B, Nickerson JA et al. The retinoblastoma gene product is a cell cycle-dependent, nuclear matrix-associated protein. Proc Natl Acad Sci USA 1994; 91(1):418-422.
97. Durfee T, Mancini MA, Jones D et al. The amino-terminal region of the retinoblastoma gene product binds a novel nuclear matrix protein that colocalizes to centers for RNA processing. J Cell Biol 1994; 127(3):609-622.
98. Markiewicz E, Dechat T, Foisner R et al. Lamin A/C binding protein LAP2alpha is required for nuclear anchorage of retinoblastoma protein. Mol Biol Cell 2002; 13(12):4401-4413.
99. Angus SP, Solomon DA, Kuschel L et al. Retinoblastoma tumor suppressor: Analyses of dynamic behavior in living cells reveal multiple modes of regulation. Mol Cell Biol 2003; 23(22):8172-8188.
100. Kennedy BK, Barbie DA, Classon M et al. Nuclear organization of DNA replication in primary mammalian cells. Genes Dev 2000; 14(22):2855-2868.
101. Dimitrova DS, Berezney R. The spatio-temporal organization of DNA replication sites is identical in primary, immortalized and transformed mammalian cells. J Cell Sci 2002; 115(Pt 21):4037-4051.
102. Orr-Weaver TL. Drosophila chorion genes: Cracking the eggshell's secrets. Bioessays 1991; 13(3):97-105.
103. Keller C, Ladenburger EM, Kremer M et al. The origin recognition complex marks a replication origin in the human TOP1 gene promoter. J Biol Chem 2002; 277(35):31430-31440.
104. Sterner JM, Dew-Knight S, Musahl C et al. Negative regulation of DNA replication by the retinoblastoma protein is mediated by its association with MCM7. Mol Cell Biol 1998; 18(5):2748-2757.
105. Gladden AB, Diehl JA. The cyclin D1-dependent kinase associates with the prereplication complex and modulates RB.MCM7 binding. J Biol Chem 2003; 278(11):9754-9760.
106. Sever-Chroneos Z, Angus SP, Fribourg AF et al. Retinoblastoma tumor suppressor protein signals through inhibition of cyclin-dependent kinase 2 activity to disrupt PCNA function in S phase. Mol Cell Biol 2001; 21(12):4032-4045.
107. Takemura M, Kitagawa T, Izuta S et al. Phosphorylated retinoblastoma protein stimulates DNA polymerase alpha. Oncogene 1997; 15(20):2483-2492.
108. Krucher NA, Zygmunt A, Mazloum N et al. Interaction of the retinoblastoma protein (pRb) with the catalytic subunit of DNA polymerase delta (p125). Oncogene 2000; 19(48):5464-5470.
109. Pennaneach V, Salles-Passador I, Munshi A et al. The large subunit of replication factor C promotes cell survival after DNA damage in an LxCxE motif- and Rb-dependent manner. Mol Cell 2001; 7(4):715-727.
110. Pennaneach V, Barbier V, Regazzoni K et al. Rb Inhibits E2F-1-induced cell death in a LXCXE-dependent manner by active repression. J Biol Chem 2004; 279(22):23376-23383.
111. Dyson N. The regulation of E2F by pRB-family proteins. Genes Dev 1998; 12(15):2245-2262.
112. Kovesdi I, Reichel R, Nevins JR. Identification of a cellular transcription factor involved in E1A trans-activation. Cell 1986; 45(2):219-228.
113. Chellappan SP, Hiebert S, Mudryj M et al. The E2F transcription factor is a cellular target for the RB protein. Cell 1991; 65(6):1053-1061.
114. Kaelin Jr WG, Krek W, Sellers WR et al. Expression cloning of a cDNA encoding a retinoblastoma-binding protein with E2F-like properties. Cell 1992; 70(2):351-364.
115. Helin K, Lees JA, Vidal M et al. A cDNA encoding a pRB-binding protein with properties of the transcription factor E2F. Cell 1992; 70(2):337-350.
116. Frolov MV, Dyson NJ. Molecular mechanisms of E2F-dependent activation and pRB-mediated repression. J Cell Sci 2004; 117(Pt 11):2173-2181.
117. Trimarchi JM, Lees JA. Sibling rivalry in the E2F family. Nat Rev Mol Cell Biol 2002; 3(1):11-20.
118. DeGregori J, Kowalik T, Nevins JR. Cellular targets for activation by the E2F1 transcription factor include DNA synthesis- and G1/S-regulatory genes. Mol Cell Biol 1995; 15(8):4215-4224.
119. DeGregori J. The genetics of the E2F family of transcription factors: Shared functions and unique roles. Biochim Biophys Acta 2002; 1602(2):131-150.
120. Muller H, Helin K. The E2F transcription factors: Key regulators of cell proliferation. Biochim Biophys Acta 2000; 1470(1):M1-12.

121. Ishida S, Huang E, Zuzan H et al. Role for E2F in control of both DNA replication and mitotic functions as revealed from DNA microarray analysis. Mol Cell Biol 2001; 21(14):4684-4699.
122. Muller H, Bracken AP, Vernell R et al. E2Fs regulate the expression of genes involved in differentiation, development, proliferation, and apoptosis. Genes Dev 2001; 15(3):267-285.
123. Polager S, Kalma Y, Berkovich E et al. E2Fs up-regulate expression of genes involved in DNA replication, DNA repair and mitosis. Oncogene 2002; 21(3):437-446.
124. Ren B, Cam H, Takahashi Y et al. E2F integrates cell cycle progression with DNA repair, replication, and G(2)/M checkpoints. Genes Dev 2002; 16(2):245-256.
125. Harbour JW, Luo RX, Dei Santi A et al. Cdk phosphorylation triggers sequential intramolecular interactions that progressively block Rb functions as cells move through G1. Cell 1999; 98(6):859-869.
126. Helin K, Harlow E, Fattaey A. Inhibition of E2F-1 transactivation by direct binding of the retinoblastoma protein. Mol Cell Biol 1993; 13(10):6501-6508.
127. Flemington EK, Speck SH, Kaelin Jr WG. E2F-1-mediated transactivation is inhibited by complex formation with the retinoblastoma susceptibility gene product. Proc Natl Acad Sci USA 1993; 90(15):6914-6918.
128. Weintraub SJ, Prater CA, Dean DC. Retinoblastoma protein switches the E2F site from positive to negative element. Nature 1992; 358(6383):259-261.
129. Weintraub SJ, Chow KN, Luo RX et al. Mechanism of active transcriptional repression by the retinoblastoma protein. Nature 1995; 375(6534):812-815.
130. Brehm A, Miska EA, McCance DJ et al. Retinoblastoma protein recruits histone deacetylase to repress transcription. Nature 1998; 391(6667):597-601.
131. Dunaief JL, Strober BE, Guha S et al. The retinoblastoma protein and BRG1 form a complex and cooperate to induce cell cycle arrest. Cell 1994; 79(1):119-130.
132. Strober BE, Dunaief JL, Guha et al. Functional interactions between the hBRM/hBRG1 transcriptional activators and the pRB family of proteins. Mol Cell Biol 1996; 16(4):1576-1583.
133. Strobeck MW, Knudsen KE, Fribourg AF et al. BRG-1 is required for RB-mediated cell cycle arrest. Proc Natl Acad Sci USA 2000; 97(14):7748-7753.
134. Zhang HS, Gavin M, Dahiya A et al. Exit from G1 and S phase of the cell cycle is regulated by repressor complexes containing HDAC-Rb-hSWI/SNF and Rb-hSWI/SNF. Cell 2000; 101(1):79-89.
135. Dahiya A, Wong S, Gonzalo S et al. Linking the Rb and polycomb pathways. Mol Cell 2001; 8(3):557-569.
136. Nielsen SJ, Schneider R, Bauer UM et al. Rb targets histone H3 methylation and HP1 to promoters. Nature 2001; 412(6846):561-565.
137. Harbour JW, Dean DC. The Rb/E2F pathway: Expanding roles and emerging paradigms. Genes Dev 2000; 14(19):2393-2409.
138. Markey MP, Angus SP, Strobeck MW et al. Unbiased analysis of RB-mediated transcriptional repression identifies novel targets and distinctions from E2F action. Cancer Res 2002; 62(22):6587-6597.
139. Qin XQ, Livingston DM, Ewen M et al. The transcription factor E2F-1 is a downstream target of RB action. Mol Cell Biol 1995; 15(2):742-755.
140. Wu L, Timmers C, Maiti B et al. The E2F1-3 transcription factors are essential for cellular proliferation. Nature 2001; 414(6862):457-462.
141. Li FX, Zhu JW, Hogan CJ et al. Defective gene expression, S phase progression, and maturation during hematopoiesis in E2F1/E2F2 mutant mice. Mol Cell Biol 2003; 23(10):3607-3622.
142. Cayirlioglu P, Ward WO, Silver Key SC et al. Transcriptional repressor functions of Drosophila E2F1 and E2F2 cooperate to inhibit genomic DNA synthesis in ovarian follicle cells. Mol Cell Biol 2003; 23(6):2123-2134.
143. Jiang W, Hunter T. Identification and characterization of a human protein kinase related to budding yeast Cdc7p. Proc Natl Acad Sci USA 1997; 94(26):14320-14325.
144. Kumagai H, Sato N, Yamada M et al. A novel growth- and cell cycle-regulated protein, ASK, activates human Cdc7-related kinase and is essential for G1/S transition in mammalian cells. Mol Cell Biol 1999; 19(7):5083-5095.
145. Jiang W, McDonald D, Hope TJ et al. Mammalian Cdc7-Dbf4 protein kinase complex is essential for initiation of DNA replication. Embo J 1999; 18(20):5703-5713.
146. Walter JC. Evidence for sequential action of cdc7 and cdk2 protein kinases during initiation of DNA replication in Xenopus egg extracts. J Biol Chem 2000; 275(50):39773-39778.
147. Angus SP, Mayhew CN, Solomon DA et al. RB reversibly inhibits DNA replication via two temporally distinct mechanisms. Mol Cell Biol 2004; 24(12):5404-5420.
148. Costanzo V, Shechter D, Lupardus PJ et al. An ATR- and Cdc7-dependent DNA damage checkpoint that inhibits initiation of DNA replication. Mol Cell 2003; 11(1):203-213.

149. Montagnoli A, Bosotti R, Villa F et al. Drf1, a novel regulatory subunit for human Cdc7 kinase. Embo J 2002; 21(12):3171-3181.
150. Yanow SK, Gold DA, Yoo HY et al. Xenopus Drf1, a regulator of Cdc7, displays checkpoint-dependent accumulation on chromatin during an S-phase arrest. J Biol Chem 2003; 278(42):41083-41092.
151. Mayhew CN, Perkin LM, Zhang X et al. Discrete signaling pathways participate in RB-dependent responses to chemotherapeutic agents. Oncogene 2004.
152. Sage J, Miller AL, Perez-Mancera PA et al. Acute mutation of retinoblastoma gene function is sufficient for cell cycle reentry. Nature 2003; 424(6945):223-228.
153. Philips A, Huet X, Plet A et al. The retinoblastoma protein is essential for cyclin A repression in quiescent cells. Oncogene 1998; 16(11):1373-1381.
154. Angus SP, Fribourg AF, Markey MP et al. Active RB elicits late G1/S inhibition. Exp Cell Res 2002; 276(2):201-213.
155. Chevalier S, Tassan JP, Cox R et al. Both cdc2 and cdk2 promote S phase initiation in Xenopus egg extracts. J Cell Sci 1995; 108(Pt 5):1831-1841.
156. Fotedar A, Cannella D, Fitzgerald P et al. Role for cyclin A-dependent kinase in DNA replication in human S phase cell extracts. J Biol Chem 1996; 271(49):31627-31637.
157. Pagano M, Pepperkok R, Verde F et al. Cyclin A is required at two points in the human cell cycle. Embo J 1992; 11(3):961-971.
158. Pagano M, Pepperkok R, Lukas J et al. Regulation of the cell cycle by the cdk2 protein kinase in cultured human fibroblasts. J Cell Biol 1993; 121(1):101-111.
159. Yan H, Newport J. An analysis of the regulation of DNA synthesis by cdk2, Cip1, and licensing factor. J Cell Biol 1995; 129(1):1-15.
160. Ohtsubo M, Theodoras AM, Schumacher J et al. Human cyclin E, a nuclear protein essential for the G1-to-S phase transition. Mol Cell Biol 1995; 15(5):2612-2624.
161. Resnitzky D, Hengst L, Reed SI. Cyclin A-associated kinase activity is rate limiting for entrance into S phase and is negatively regulated in G1 by p27Kip1. Mol Cell Biol 1995; 15(8):4347-4352.
162. Coverley D, Laman H, Laskey RA. Distinct roles for cyclins E and A during DNA replication complex assembly and activation. Nat Cell Biol 2002; 4(7):523-528.
163. Ortega S, Prieto I, Odajima J et al. Cyclin-dependent kinase 2 is essential for meiosis but not for mitotic cell division in mice. Nat Genet 2003; 35(1):25-31.
164. Tetsu O, McCormick F. Proliferation of cancer cells despite CDK2 inhibition. Cancer Cell 2003; 3(3):233-245.
165. DeGregori J, Leone G, Ohtani K et al. E2F-1 accumulation bypasses a G1 arrest resulting from the inhibition of G1 cyclin-dependent kinase activity. Genes Dev 1995; 9(23):2873-2887.
166. Geng Y, Yu Q, Sicinska E et al. Cyclin E ablation in the mouse. Cell 2003; 114(4):431-443.
167. Walker DH, Maller JL. Role for cyclin A in the dependence of mitosis on completion of DNA replication. Nature 1991; 354(6351):314-317.
168. Knudsen KE, Fribourg AF, Strobeck MW et al. Cyclin A is a functional target of retinoblastoma tumor suppressor protein-mediated cell cycle arrest. J Biol Chem 1999; 274(39):27632-27641.
169. Fang F, Newport JW. Distinct roles of cdk2 and cdc2 in RP-A phosphorylation during the cell cycle. J Cell Sci 1993; 106(Pt 3):983-994.
170. Voitenleitner C, Fanning E, Nasheuer HP. Phosphorylation of DNA polymerase alpha-primase by cyclin A-dependent kinases regulates initiation of DNA replication in vitro. Oncogene 1997; 14(13):1611-1615.
171. Bashir T, Horlein R, Rommelaere J et al. Cyclin A activates the DNA polymerase delta -dependent elongation machinery in vitro: A parvovirus DNA replication model. Proc Natl Acad Sci USA 2000; 97(10):5522-5527.
172. Almasan A, Yin Y, Kelly RE et al. Deficiency of retinoblastoma protein leads to inappropriate S-phase entry, activation of E2F-responsive genes, and apoptosis. Proc Natl Acad Sci USA 1995; 92(12):5436-5440.
173. Angus SP, Wheeler LJ, Ranmal SA et al. Retinoblastoma tumor suppressor targets dNTP metabolism to regulate DNA replication. J Biol Chem 2002; 277(46):44376-44384.
174. Chabes AL, Pfleger CM, Kirschner MW et al. Mouse ribonucleotide reductase R2 protein: A new target for anaphase-promoting complex-Cdh1-mediated proteolysis. Proc Natl Acad Sci USA 2003; 100(7):3925-3929.
175. Chabes AL, Bjorklund S, Thelander L. S Phase-specific transcription of the mouse ribonucleotide reductase R2 gene requires both a proximal repressive E2F-binding site and an upstream promoter activating region. J Biol Chem 2004; 279(11):10796-10807.
176. Tanaka H, Arakawa H, Yamaguchi T et al. A ribonucleotide reductase gene involved in a p53-dependent cell-cycle checkpoint for DNA damage. Nature 2000; 404(6773):42-49.
177. Nakano K, Balint E, Ashcroft M et al. A ribonucleotide reductase gene is a transcriptional target of p53 and p73. Oncogene 2000; 19(37):4283-4289.

Chapter 4

New Insights into Transcriptional Regulation by Rb:

One Size No Longer Fits All

Peggy J. Farnham*

Abstract

The retinoblastoma (Rb) protein is a key regulator of cell proliferation, differentiation, and tumorigenesis. Initial studies of Rb revealed that it binds to, and decreases the activity of, the E2F family of transcription factors. Over the last decade, the mechanisms by which Rb regulates E2F activity have been well-studied. These investigations have lead to a commonly held belief that Rb functions solely as a transcriptional repressor. However, although not as commonly discussed, there are many examples of Rb synergizing with site-specific transcription factors to activate transcription. This Rb-mediated activation appears to be cell type-specific, transcription factor-specific, and even promoter-specific. This chapter details some of the examples of Rb-mediated transcriptional activation and suggests future studies that could provide insight into the mechanisms by which Rb can function to positively regulate transcription.

The Classic Model for Rb Function

The retinoblastoma (Rb) gene was the first tumor suppressor gene cloned[1-4] and hundreds of investigators have since studied the biological consequences of deregulation of Rb in human tumors, in mouse models, and in cell culture. The results of these studies have been used to generate a commonly held model for the mechanism by which Rb modulates cell proliferation. In brief, Rb is proposed to block cell cycle progression and to promote differentiation by negatively regulating the transcription of genes whose products are required for the G1/ S phase transition and DNA replication. It has been postulated that Rb-mediated repression is due to its ability to interact with proteins whose biochemical activities favor the creation of an inactive chromatin structure.[5] For example, several groups have found that Rb can interact with a histone deacetylase to repress transcription through the removal of acetyl groups from the lysine tails of histones. The removal of this post-translational modification is associated with a compacted (inactive) nucleosomal structure, which leads to reduced transcriptional competence.[6-8] Rb can also interact with SUV39H1, a human histone methyltransferase that specifically methylates histone H3 at lysine 9, resulting in the formation of heterochromatin and transcriptional repression.[9] Another activity with which Rb interacts is human SWI/SNF, a protein complex which mediates transcriptional repression via an ATP-dependent remodeling of chromatin. Finally, Rb can cooperate with DNA methyltransferase 1 to repress promoters

*Peggy J. Farnham—McArdle Laboratory for Cancer Research, University of Wisconsin Medical School, 1400 University Avenue, Madison, Wisconsin 53706, U.S.A.
Email: farnham@oncology.wisc.edu

Rb and Tumorigenesis, edited by Maurizio Fanciulli. ©2006 Eurekah.com and Springer Business+Science Media.

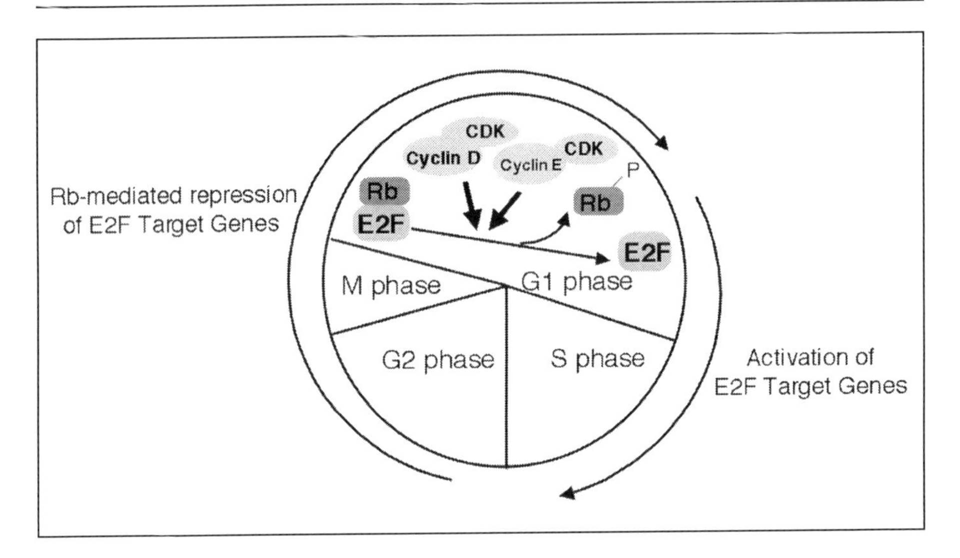

Figure 1. The classic model of Rb-mediated transcriptional regulation of E2F target genes. Rb serves as a transcriptional repressor of E2F target genes in G0 phase cells. The mechanism by which Rb represses transcription involves the ability of Rb to bring proteins such as histone methyltransferases or histone deacetylases to the basal promoter region. The interaction between Rb and E2F is weakened by the action of cyclin-dependent kinases that phosphorylate Rb in a cell cycle position-dependent manner. Therefore, in late G1 and during S phase, E2F is not bound by Rb and E2F target genes are transcribed.

containing E2F sites. In this case, it is thought that methylation of CpG dinucleotides in the promoter region by DNA methyltransferase 1 leads to the interaction of methyl-CpG binding proteins with the DNA, followed by the recruitment of chromatin remodeling enzymes such as SUV39H1.

Because Rb lacks a DNA binding domain, it must be localized to promoters through interactions with site-specific DNA binding proteins.[10] The first transcription factor shown to interact with Rb was E2F1.[11-13] Subsequently, it has been shown that Rb can bind to several members of the E2F family, with the highest affinity towards E2F1, E2F2, and E2F3 and the lowest affinity towards E2F4. The interaction between Rb and the E2Fs is regulated by cell cycle-dependent phosphorylation of Rb (Fig. 1). During G0 and G1 phases of the cell cycle, hypophosphorylated Rb binds to and inhibits the transcriptional activity of E2F-regulated promoters. As cells progress through G1, cyclin D/cdk4, cyclin D/cdk6, and cyclin E/cdk2 complexes are sequentially activated and then phosphorylate Rb. Hyperphosphorylation of Rb results in its dissociation from, and the resultant activation of, E2F complexes during late G1.[14] Since the genes regulated by E2F family members mediate cell cycle progression, DNA repair, DNA replication and DNA recombination, Rb can participate in the control of these processes through its interaction with, and functional inhibition of, the E2F family members.[15]

Confounding Facts About Rb Function

The model for Rb function described above is entirely consistent with its role as a tumor suppressor; i.e., the presence of Rb normally keeps E2F-regulated genes under tight control, whereas loss of Rb in tumors allows such genes to be transcribed at inappropriately high levels, leading to enhanced proliferation and neoplastic transformation. However, a growing body of evidence suggests that the role of Rb in the cell may be more complex. For example, Rb is not always lost or mutated in tumors. In fact, Rb has been shown to be present at increased levels in colon tumors, as compared to normal colon tissue.[16] One explanation of this seemingly paradoxical finding is that the increased Rb may play a role in the prevention of apoptosis. It has been shown that E2Fs, when overexpressed, can activate apoptosis-inducing genes.[17] Several signal transduction pathways that are often inappropriately activated in cancers have been implicated in controlling the levels and/or activity of the E2F family members. Thus, it is possible that certain colon tumors arise due to initial mutations which activate signaling pathways upstream of E2F, leading to a higher than normal amount of active E2F in the cell. If so, then loss of Rb would allow even higher E2F activity, perhaps resulting in the increased E2F-mediated transcriptional activation of pro-apoptotic genes. Aberrant expression of Rb in a tumor may allow cells to survive long enough to acquire functional mutations in genes that mediate apoptosis. Considered in this light, retaining or increasing levels of Rb may be critical for tumor development, whereas loss of Rb would be detrimental to tumor formation. Accordingly, Yamamoto et al have shown that introduction of antisense mRNA to Rb results in apotosis in colon cancer cells.[16] While these findings support the classical model of Rb functioning to repress E2F-regulated genes, it is possible that the increased Rb in colon tumors may play an alternate role. For example, others have shown that expression of a constitutively active Rb in mouse mammary glands results in the development of hyperplastic nodules and adenocarcinomas.[18] In this case, Rb appears to be promoting tumor formation when overexpressed in otherwise normal cells (i.e., cells that do not have excessive E2F activity). These results suggest that Rb may play a role in tumor initiation in mammary cells through a mechanism distinct from repressing transcription of E2F target genes.

Another observation about Rb that does not fit the commonly held model is that Rb is not always released from the chromatin in S phase.[19] Using a relatively unbiased approach that relies upon a combination of chromatin immunoprecipitation and CpG microarray analysis, Wells et al identified genomic sites bound by Rb in Raji cells. As expected, characterization of these sites identified a subset that showed the cell cycle-dependent changes in protein-DNA occupancy that would be predicted by the classic model. For example, some of the newly identified sites were bound by Rb and E2F in G0 phase, but Rb could not be detected on the site in S phase. On the other hand, certain promoters showed an increase in Rb recruitment in S phase, whereas others showed high level, constitutive binding of Rb throughout the cell cycle. These surprising findings raise the question as to whether Rb mediates the same biochemical activities in G0 and in S phase. In other words, is Rb always a transcriptional repressor?

It is intriguing that one of the sites that showed an increased amount of bound Rb in S phase was the promoter for the nuclear oncogene c-Myc. The expression of c-Myc correlates with cell proliferation and the *c-Myc* promoter displays robust activity in S phase. However, too much c-Myc can lead to neoplastic transformation and, therefore, levels of c-Myc must be tightly controlled.[20,21] One interpretation of the chromatin immunoprecipitation data of Wells et al[19] is that Rb may function as a rheostat. In this model, Rb would keep the Myc gene in a fully off position in quiescent cells but would work in opposition to S-phase specific activators to keep c-Myc transcription at submaximal levels in S phase. However, an alterative explanation could be that Rb is involved in activation of the *c-Myc* promoter in S phase, in addition to its role as a repressor in G0 phase. In other words, Rb may function as a switch hitter, exchanging corepressors in G0 phase with coactivators in S phase. The concept that Rb may promote tumor formation and/or proliferation by activating transcription of oncogenes such as c-Myc is tantalizing. However, to date, the consequences of removing Rb from the *c-Myc* promoter in S phase has not been determined. Therefore, it is not possible to conclude that Rb is an S phase-specific activator of the *c-Myc* promoter. Although the exact role that Rb played in the regulation of the promoters to which it was bound in S phase was not determined in the studies of Wells et al,[19] there is evidence, as described in the following section, to support the hypothesis that Rb can serve as a transcriptional activator at certain promoters.

Evidence in Support of the Role of Rb as a Transcriptional Activator

The first protein shown to cooperate with Rb to stimulate transcription was the site-specific transcription factor Sp1. Robbins et al identified a cis element that mediates transcriptional activation in response to Rb; this element was shown to bind Sp1.[22,23] Several other genes containing Sp1 sites have also been shown to be activated by Rb, including the fourth promoter of the *insulin like growth factor II* gene and the hamster *dihydrofolate reductase* (*dhfr*) gene.[24,25] The exact mechanism by which Rb stimulates Sp1-mediated transcription is unclear in many cases and the ability of Rb to function as an activator seems to be cell type-dependent and promoter-specific.[26] In some cases, it has been suggested that Rb directly interacts with Sp1 to stimulate Sp1 transcriptional activity,[25] whereas the action of Rb on Sp1 activity has been proposed to be indirect in other cases. For example, it has been proposed that Rb can stimulate transcription by interfering with the interaction between Sp1 and a negative regulator of Sp1 activity.[27] Interestingly, Johnson-Pais et al[28] showed that Sp1 activity can be inhibited by physical interaction with mdm2 and that expression of Rb results in the release of mdm2 from Sp1, most likely through Rb sequestering the mdm2 protein. Clearly, the degree to which Rb could stimulate Sp1 activity would be dependent on the comparative levels of Sp1 and mdm2 in a particular cell.

Rb has also been proposed to function as an activator in other transcription complexes. For example, Thomas et al[29] have shown that Rb physically interacts with the site-specific transcription factor CBFA1 to activate an osteoblast-specific reporter. CBFA1 is a transcriptional regulator that is critical for inducing osteoblast differentation. Thus, loss of Rb may lead to decreased CBFA1 transcriptional activity and a resultant defect in differentiation. The inability to achieve a differentiated state may lead to inappropriate proliferation and, eventually, to tumorigenesis. Interestingly, osteosarcoma is the second most common tumor after retinoblastoma itself among individuals with inherited heterozygous loss of the *Rb* gene.[30] In addition, loss of Rb occurs in up to 60% of sporadic osteosarcomas.[31,32] These findings support the hypothesis that loss of Rb alters the balance between proliferation and differentiation in osteoblasts. Thomas et al performed chromatin immunoprecipitation assays to demonstrate that Rb is recruited to the osteocalcin and osteopontin promoters, both of which are regulated by CBFA1. However, the mechanism by which Rb stimulates transcriptional activation when bound to these promoters is unknown. For example, does Rb enhance CBFA1-mediated recruitment of basal transcription factors or alternatively does an CBFA-1/Rb complex help recruit another site-specific transcription factor?

The Jun family of transcription factors has also been implicated in Rb-mediated transcriptional activation. Xin et al[33] showed that Rb positively regulates expression of the *p202* gene, an interferon- and differentiation-inducible phosphoprotein. Initial experiments demonstrated that an AP-1 site (to which Jun family members bind in cooperation with Fos family members) is critical for Rb-mediated transcriptional activation of the *p202* gene. The authors then showed that Rb cooperates with JunD to provide an even greater transcriptional activation of the *p202* promoter. These studies did not characterize the mechanism by which Rb cooperates with JunD. Therefore, it is not yet known if Rb is bound to the *p202* promoter region (as in the case with CBFA1) or if Rb works via removal of an inhibitor (as in certain cases of Sp1-mediated transcription). Others have also shown that Rb can cooperate with Jun family members. Slack et al found that Rb can cooperate with c-Jun to activate the *DNA methyltransferase 1* (*dnmt1*) promoter by facilitating the in vitro binding of c-Jun to a noncanonical AP1 site.[34] However, Rb could not be detected in the DNA-bound protein complex. The authors offer two alternative explanations for these results. First, Rb may enhance c-Jun binding to DNA by removal of an inhibitor of c-Jun DNA binding activity (similar to the Sp1 situation). Alternatively, Rb may initially bind to the DNA with c-Jun but be lost from the DNA-bound complex during the electrophoresis of the gel shift assay. Interestingly, although Rb can be detected (along with CBFA1) at the *osteopontin* promoter using in vivo assays, Thomas et al[29] were not able to detect Rb in the CBFA1-DNA complex in vitro. Results such as these point out the advantages of

Table 1. Factors which synergize with Rb to activate transcription

Transcription Factors*	References
Sp1	23-28
CBFA1	29
Jun family	33-34
C/EBP family	35-36
AP-2	38-39
Myo D and related factors	40-41
ATF family	42-43
AH receptor	44
NF-κB	45
Nuclear receptors	46-49

*Note that the list is not comprehensive, but rather is meant to provide insight into the many different types of factors that can functionally interact with Rb.

using in vivo methods such as chromatin immunoprecipitation to analyze protein-DNA interactions, rather than relying on in vitro assays.

Another family of transcription factors with which Rb cooperates to activate transcription includes the C-EBP proteins. For example, Charles et al[35] revealed that Rb stimulates transcription of the *surfactant protein D* promoter via a C-EBP binding motif. They also showed that Rb can physically interact with C-EBP alpha, C-EBP beta and C-EBP delta and can be detected in C-EBP/DNA complexes in vitro. Similarly, Gery et al[36] have shown that Rb, in the presence of C-EBP beta but not alone, activates promoters of myeloid specific genes such as the *granulocyte colony-stimulating factor receptor* (*G-CSFR*) and *mim1*. In contrast, other studies have shown that Rb can decrease C-EBP beta DNA binding activity, resulting in inhibition of C-EBP beta-mediated transcription in preadipocytes.[37] Thus, similar to the effects on Sp1 activity, the direction in which Rb influences transcription mediated by C-EBP family members appears to be cell type- and promoter-specific. It is likely that the presence of other site-specific DNA binding factors bound to a particular promoter will determine the exact role that Rb plays in transcriptional regulation of C-EBP target genes.

Several groups have also reported an interaction between Rb and AP-2. For example, Batsche et al[38] showed that Rb can physically interact with AP-2 in vitro and in vivo and that it cooperates with AP-2 to activate the *E-cadherin* promoter. Similarly, Decary et al[39] have shown that Rb activates the *bcl-2* promoter through an AP-2 site. Interestingly, both of these studies showed that the synergy between Rb and AP2 is cell type-dependent, occurring in epithelial cells but not in fibroblasts. Importantly, Decary et al use chromatin immunoprecipitation assays to show that Rb can bind to the *bcl-2* and *E-cadherin* promoters.

The cases described above in which Rb stimulates transcription do not constitute a comprehensive list of all known Rb-activated promoters (Table 1). Rather, they simply provide insight concerning the different types of transcription factors, target genes, and cell types for which Rb may serve a role other than as a transcriptional repressor. There are many other reported instances of Rb-mediated transcriptional activation. For example, Rb has been shown to synergize with transcriptional activators involved in muscle differentiation such as MyoD, Myogenins and Myf-5,[40,41] ATFa and ATF2,[42,43] the AH receptor[44] and NF-kb.[45] Rb has also been shown to cooperate with several different steroid receptors to activate transcription. For example, Singh et al[46,47] showed that hBrm and Rb (but not the related proteins p107 or p130) cooperated to activate glucocorticoid-receptor mediated transcription. Also, Balasenthil et al[48] showed that Rb positively regulates the cyclin D1 promoter via interactions with the estrogen receptor

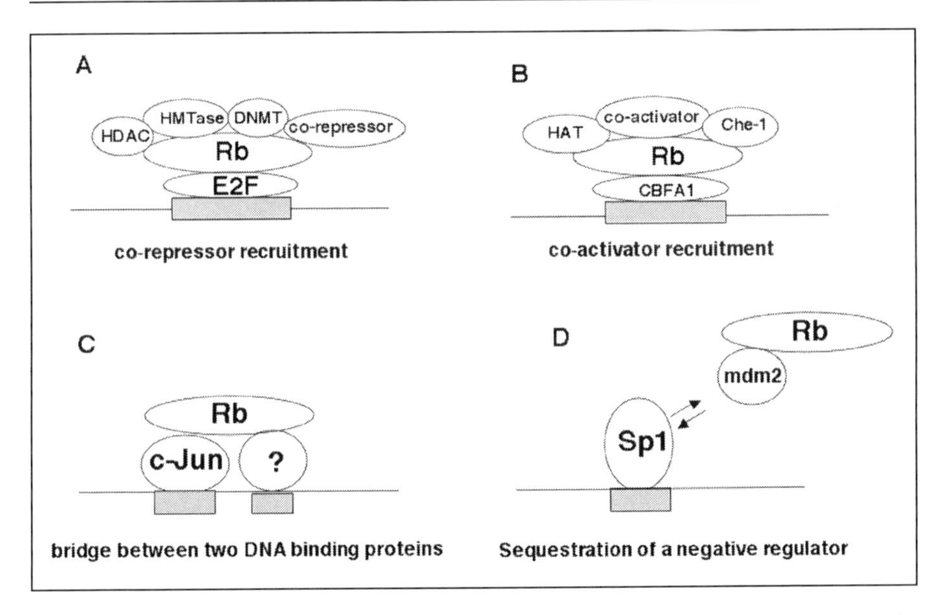

Figure 2. Mechanisms by which Rb can influence transcription. Depiction of four different mechanisms by which Rb may influence transcription. In panel A, Rb represses transcription of E2F target genes by serving as a platform for the binding of histone methyltransferases (HMTases), histone deacetylases (HDACs), DNA methyltransferases (DNMTs), or other corepressors. Similarly, in panel B, Rb activates transcription of CBFA1 target genes by serving as a platform for the binding of histone acetyltransferases (HATs) or other coactivators such as Che-1. In panel C, Rb stimulates transcription by serving as a bridge between two DNA binding factors and enhances their binding to the promoter DNA. Finally, in panel D, Rb sequesters a factor which can inhibit either the DNA binding or activity of a transcription factor (such as Sp1).

coactivator PELP1/MNAR. Finally, Hofman et al[49] demonstrate that Rb serves as a coactivator of androgen receptor-mediated transcription.

In summary, a growing body of evidence supports the hypothesis that Rb can function to mediate either activation or repression of transcription, most likely by serving as a platform for the recruitment of both coactivators and corepressors. As noted above, the mechanisms by which Rb serves as a repressor of E2F-activated transcription have been fairly well-studied (Fig. 2A). For example, Rb can interact with SUV39H1, a methylase specific for lysine 9 of histone H3. Nicolas et al[50] have shown, using chromatin immunoprecipitation and synchronized cell populations, that Rb-repressed promoters such as *dhfr* show higher levels of histone H3 methylated on lysine 9 in G0 phase (when Rb is bound) than in S phase (when Rb is released). They also showed that levels of histone H3 acetylated on lysine 9 increase at the *dhfr* promoter as cells progress from G0 to S phase. Others have shown that HDAC1 is associated with the *dhfr* promoter in G0 phase, but not in S phase.[51] Thus, a model can be developed for G0 phase-specific Rb-mediated repression of the *dhfr* promoter that entails Rb-mediated recruitment of an HDAC (to deacetylate H3), followed by or concomitant with Rb-mediated recruitment of SUV39H1 for methylation of lysine 9 of histone H3. For many promoters, the phosphorylation of Rb in S phase would result in disruption of Rb/E2F interactions, loss of Rb (and thus loss of HDACs and histone methyltransferases) from the promoter, and activation of transcription. However, it must be recalled that many promoters, such as the *c-Myc* promoter, are transcriptionally active and bound by Rb in S phase. Recent studies suggest a mechanism by which Rb might change roles and serve as a transcriptional activator in S phase (Fig. 2B). Fanciulli et al showed that a protein called Che-1 can reverse the Rb-mediated repression of the *dhfr* promoter in Saos-2 cells, but has no effect on basal transcription or E2F1-mediated activation of the *dhfr* promoter in the absence of Rb.[52] Further studies revealed that Che-1 can bind directly to and displace HDAC1 from Rb.[53] The binding of Che-1 to E2F target promoters is cell cycle regulated, showing the highest binding in S phase. Interestingly, Che-1 can also interact with the RPB11 subunit of RNAP II. Thus, the cell cycle-dependent interaction of Che-1 with Rb could turn Rb from a repressor to an activator, by displacing HDACs and serving as an adaptor between Rb and the basal transcriptional machinery.[52]

Because Rb has been shown to bind more than 100 proteins, many of which are transcriptional regulators,[10] it would not be surprising to find that the ability to repress E2F activity is an important, but not sole, function of Rb, especially in differentiated cell types. Unfortunately, very few studies have been performed using chromatin IP to analyze markers of active vs inactive chromatin on non E2F-regulated promoters that are activated by Rb. However, one fairly recent study did use chromatin IP to show a correlation between transcriptional activation, binding of Rb and AP2, and the presence of acetylated histone H4 (a mark of active chromatin) on the *bcl-2* promoter.[39] This suggests that perhaps Rb can recruit a histone acetyltransferase (Fig. 2b). As described above, it is also possible that Rb mediates transcriptional activation via mechanisms that are independent of changes in chromatin structure, such as assisting in the recruitment of other site-specific factors (Fig. 2C) or the sequestration of inhibitory proteins (Fig. 2D).

Future Studies

Clearly, the hypothesis that Rb can serve as a transcriptional activator on certain promoters, often in a cell-type specific manner, is supported by a number of different experiments. However, further investigation is required to (a) determine the relative contribution of Rb as an activator vs. a repressor and (b) to determine the mechanism(s) by which Rb mediates activation. Both points can be addressed by performing ChIP-chip studies using global genomic arrays and antibodies to Rb, markers of active vs. inactive chromatin, and components of the basal transcriptional machinery. In addition, it will be very interesting to compare the binding patterns of Rb, site-specific factors shown to interact with Rb (e.g., E2Fs, AP2, CBFA1, Jun family members, Sp1, and steroid receptors), and chromatin modifying proteins in different cell types (e.g., epithelial vs. fibroblasts or tumor vs. normal cells). Finally, it will be critical to specifically remove Rb from a cell using siRNA technology and then determine the effect of this removal on the chromatin state and/or on the recruitment of other transcription factors to a particular promoter. Results from such studies will likely enforce the conclusion that Rb does not play the same role at all loci but rather serves a unique role at each promoter that it regulates.

Acknowledgements

This work was supported in part by Public Health Service grant CA45240. We thank Bill Schelman for critical reading of the manuscript.

References

1. Fung YK, Murphree AL, T'Ang A et al. Structural evidence for the authenticity of the human retinoblastoma gene. Science 1987; 236:1657-1661.
2. Lee W-H, Bookstein R, Hong F et al. Human retinoblastoma gene: Cloning, identification, and sequence. Science 1987; 235:1394-1399.
3. Friend SH, Bernards R, Rogelj S et al. A human DNA segment with properties of the gene that predispposes to retinoblastoma and osteosarcoma. Nature 1986; 323:643-646.
4. Knudson AGJ. Mutation and cancer: Statistical study of retinoblastoma. Proc Natl Acad Sci USA 1971; 1971:820-823.
5. Zhang HS, Dean DC. Rb-mediated chromatin structure regulation and transcriptional repression. Oncogene 2001; 20:3134-3138.
6. Brehm A, Miska EA, McCance DJ et al. Retinoblastoma protein recruits histone deacetylase to repress transcription. Nature 1998; 391:597-601.
7. Luo RX, Postigo AA, Dean DC. Rb Interacts with Histone Deacetylase to repress transcription. Cell 1998; 92:463-473.

8. Magnaghi-Jaulin L, Groisman R, Naguibneva I et al. Retinoblastoma protein represses transcription by recruiting a histone deacetylase. Nature 1998; 391:601-605.
9. Neilsen SJ, Schneider R, Bauer U-M et al. Rb targets histone H3 methylation and HP1 to promoters. Nature 2001; 412:561-565.
10. Morris EJ, Dyson NJ. Retinoblastoma protein partners. Advances in Cancer Research 2001; 82:1-54.
11. Helin K, Lees JA, Vidal M et al. A cDNA encoding a pRB-binding protein with properties of the transcription factor E2F. Cell 1992; 70:337-350.
12. Kaelin Jr WG, Krek W, Sellers WR et al. Expression cloning of a cDNA encoding a retinoblastoma-binding protein with E2F-like properties. Cell 1992; 70:351-364.
13. Shan B, Zhu X, Chen P-L et al. Molecular cloning of cellular genes encoding retinoblastoma-associated proteins: Identification of a gene with properties of the transcription factor E2F. Mol Cell Biol 1992; 12(12):5620-5631.
14. Trimarchi JM, Lees JA. Sibling rivalry in the E2F family. Nat Rev Mol Cell Biol 2002; 3:11-20.
15. Lipinski MM, Jacks T. The retinoblastoma gene family in differentiation and development. Oncogene 1999; 18:7873-7882.
16. Yamamoto H, Soh JW, Monden T et al. Paradoxical increase in retinoblastoma protein in colorectal carcinomas may protect cells from apoptosis. Clinical Cancer Res 1999; 5:1805-1815.
17. Ginsberg D. E2F1 pathways to apoptosis. FEBS Lett 2002; 529:122-125.
18. Jiang Z, Zacksenhaus E. Activation of retinoblastoma protein in mammary glands leads to ductal growth suppression, precocious differentiation, and adenocarcinoma. J Cell Biol 2002; 156:185-198.
19. Wells J, Yan PS, Cechvala M et al. Identification of novel pRb binding sites using CpG microarrays suggests that E2F recruits pRb to specific genomic sties during S phase. Oncogene 2003; 22:1445-1460.
20. Oster SK, Ho CSW, Soucie EL et al. The myc oncogene: Marvelously complex. Adv Cancer Res 2002; 84:81-154.
21. Nilsson JA, Cleveland JL. Myc pathways provoking cell suicide and cancer. Oncogene 2003; 22:9007-9021.
22. Robbins P, Horowitz J, Mulligan R. Negative regulation of human c-fos expression by the retinoblastoma gene product. Nature 1990; 346(16):668-671.
23. Udvadia AJ, Rogers KT, Higgins PDR et al. Sp-1 binds promoter elements regulated by the RB protein and Sp-1-mediated transcription is stimulated by RB coexpression. Proc Natl Acad Sci USA 1993; 90:3265-3269.
24. Kim S-J, Onwuta US, Lee YI et al. The retinoblastoma gene product regulates Sp1-mediated transcription. Mol Cell Biol 1992; 12:2455-2463.
25. Noe V, Alemany C, Chasin LA et al. Retinoblastoma protein associates with Sp1 and activates the hamster dihydrofolate reductase promoter. Oncogene 1998; 16:1931-1938.
26. Decesse JT, Medjkane S, Datto MB et al. RB regulates transcription of the p21/WAF1/CIP1 gene. Oncogene 2001; 20:962-971.
27. Chen LI, Nishinaka T, Kwan K et al. The retinoblastoma gene product RB stimulates Sp1-ediated transcription by liberating Sp1 from a negative regulator. Mol Cell Biol 1994; 14:4380-4389.
28. Johnson-Pais T, Degnin C, Thayer MJ. pRB induces Sp1 activity by relieving inhibition mediated by MDM2. Proc Natl Acad Sci USA 2001; 98:2211-2216.
29. Thomas DM, Carty SA, Piscopo DM et al. The retinoblastoma protein acts as a transcriptional coactivator required for osteogenic differentiation. Molecular Cell 2001; 8:303-316.
30. Gurney JG, Severson RK, Davis S et al. Incidence of cancer in children in the United States. Sex-, race-, and 1-year age-specific rates by histologic type. Cancer 1995; 75:2186-2195.
31. Hansen MF, Koufos A, Gallie BO et al. Osteosarcoma and retinoblastoma: A shared chromosomal mechanism revealing recessive predisposition. Proc Natl Acad Sci USA 1985; 82:6216-6220.
32. Wadayama B, Toguchida J, Shimizu T et al. Mutation spectrum of the retinoblastoma gene in osteosarcomas. Cancer Res 1994; 54:3042-3048.
33. Xin H, Pramanik R, Choubey D. Retinoblastoma (Rb) protein upregulates expression of the Ifi202 gene encoding an interferon-inducible negative regulator of cell growth. Oncogene 2003; 22:4775-4785.
34. Slack A, Pinard M, Araujo FD et al. A novel regulatory element in the dnmt1 gene that responds to coactivation by Rb and c-Jun. Gene 2001; 268:87-96.
35. Charles A, Tang X, Crouch E et al. Retinoblastoma protein complexes with C/EBP proteins and activates C/EBP-mediated transcription. J Cell Biochem 2001; 83:414-425.
36. Gery S, Gombart AF, Fung YK et al. C/EBPe interacts with retinoblastoma and E2F1 during granulopoiesis. Blood 2004; 103:828-835.
37. Cole KA, Harmon AW, Harp JB et al. Rb regulates C/EBPb-DNA-binding activity during 3T3-L1 adippogenesis. Am J Physiol Cell Physiol 2004; 286:C349-C354.

38. Batsche E, Muchardt C, Behrens J et al. RB and c-Myc activate expression of the E-cadherin gene in epithelial cells through interaction with transcription factor AP-2. Mol Cell Biol 1998; 18:3647-3658.
39. Decary S, Decesse JT, Ogryzko V et al. The retinoblastoma protein binds the promoter of the survival gene bcl-2 and regulates its transcription in epithelial cells through transcription factor AP-2. Mol Cell Biol 2002; 22:7877-7888.
40. Gu W, Schneider JW, Condorelli G et al. Interaction of myogenic factors and the retinoblastoma protein mediates muscle cell commitment and differentiationq. Cell 1993; 72:309-324.
41. Novitch BG, Spicer DB, Kim PS et al. pRb is required for MEF2-dependent gene expression as well as cell-cycle arrest during skeletal muscle differentiation. Curr Biol 1991; 9:449-459.
42. Gong Q, Huang Z, Wicks WD. Interaction of retinoblastoma gene product with transcription factors ATFa and ATF2. Arch Biochem Biophys 1995; 319:445-450.
43. Li H, Wicks WD. Retinoblastoma protein interacts with ATF2 andJNK/p38 in stimulating the transforming grwoth factor-beta2 promoter. Arch Biochem Biophys 2001; 394:1-12.
44. Elferink CJ, Ge NL, Levine A. Maximal aryhydrocarbon receptor activity depends on an interaction with the retinoblastoma protein. Mol Pharmacol 2001; 59:664-673.
45. Takebayashi T, Higashi H, Sudo H et al. NF-kB-dependent induction of cyclin D1 by retinoblastoma protein (pRB) family proteins and tumor-derived pRB mutants. J Biol Chem 2003; 278:14897-14905.
46. Singh P, Coe J, Hong W. A role for retinoblastoma protein in potentiating transcriptional activation by the glucocorticoid receptor. Nature 1995; 374:562-565.
47. Singh P, Chan SW, Hong W. Retinoblastoma protein is functionally distinct from its homologues in affecting glucocorticoid receptor-mediated transcription and apoptosis. J Biol Chem 2001; 276:13762-13770.
48. Balasenthil S, Vadlamudi RK. Functional interactions between the estrogen receptor coactivator PELP1/MNAR and retinoblastoma protein. J Biol Chem 2003; 278:22119-22127.
49. Hofman K, Swinnen JV, Claessens F et al. The retinoblastoma protein-associated transcriptional repressor RBaK interacts with the androgen receptor and enhances its transcriptional activity. J Mol Endocrinol 2003; 31:583-596.
50. Nicolas E, Roumillac C, Trouche D. Balance between acetylation and methylation of histone H3 lysine 9 on the E2F-responsive dihydrofolate reductase promoter. Mol Cell Biol 2003; 23(5):1614-1622.
51. Ferreira R, Naguibneva I, Mathieu S et al. Cell cycle-dependent recruitment of HDAC-1 correlates with deacetylation of histone H4 on an Rb-E2F target promoter. EMBO Rep 2001; 2:794-799.
52. Fanciulli M, Bruno T, Di Padova M et al. Identification of a novel partner of RNA polymerase II subunit 11, Che-1, which interacts with and affects the growth suppression function of Rb. FASEB J 2000; 14:904-912.
53. Bruno T, De Angelis R, De Nicola F et al. Che-1 affects cell growth by interfering with the recruitment of HDAC1 by Rb. Cancer Cell 2002; 2:387-399.

CHAPTER 5

Regulation of Rb Function by Noncyclin Dependent Kinases

Jaya Padmanabhan and Srikumar P. Chellappan*

Abstract

Inactivation of the retinoblastoma tumor suppressor protein, Rb, is necessary for the normal progression of the mammalian cell cycle.[1] Studies over the past fifteen years have shown that Rb protein is inactivated by kinases associated with cyclins, especially cyclins D and E, which facilitate the entry of cells from the G1 to S phase.[1-3] Though the cyclin/cdk mediated inactivation of Rb has been well studied, the role of other kinases in regulating Rb function is relatively less understood.[4] It has been shown that components of the MAP kinase cascade, including ERK kinases as well as the Raf-1 kinase can phosphorylate Rb efficiently in response to proliferative signals.[5,6] A physiological role for Raf-1 in inactivating Rb during cell cycle progression has been established.[7] These kinases seem to work in conjunction with the cyclin-cdks, facilitating phosphorylation by the latter; at the same time, over-expression of Raf-1 could inactivate Rb as efficiently as cyclin-cdks.[6,7] Similarly, it has been shown that Rb is inactivated upon apoptotic signaling as well.[8-10] Such inactivation events appear to be mediated by the p38 kinase in a human T-cell leukemia system as well as a neuronal system.[11,12] The inactivation of Rb upon apoptotic signaling seems to be totally independent of cyclins and cdks and occurring on different sites on the Rb protein.[13] In addition to the p38 kinase, the stress-induced kinase JNK1 has been shown to affect Rb and E2F functions in certain apoptotic situations.[11,14] Recently, the apoptosis signal regulating kinase, ASK1, was found to interact with Rb and overcome its anti-apoptotic activities.[15] These studies suggest that while cyclin dependent kinases are the predominant regulators of Rb, especially during cell cycle progression, other kinases are capable of functionally inactivating Rb in response to multiple stimuli. Since inactivation of the Rb protein is widespread in a wide array of human tumors,[10,16,17] understanding the mechanisms that inactivate Rb in response to normal physiological stimuli would be valuable in developing novel therapeutic strategies to combat cancer.

Introduction

The retinoblastoma tumor suppressor protein is thought to be the major regulator of mammalian cell cycle, posing a barrier at the G1/S boundary.[18-20] Rb achieves this control by repressing the transcriptional activity of the E2F family of transcription factors, mainly E2Fs 1-3.[21-25] Since many of the cellular genes necessary for DNA synthesis as well as cell cycle progression require E2F for their expression, inhibiting E2F activity appears to be an effective way to arrest cell cycle progression.[26-29] Transcriptional repression by Rb is mediated mainly through the recruitment of various corepressors, including but not limited to HDAC1,[30,31]

*Corresponding Author: Srikumar P. Chellappan—Drug Discovery Program, Department of Interdisciplinary Oncology, H. Lee Moffitt Cancer Center and Research Institute, 12902 Magnolia Drive, Tampa, Florida 33612, U.S.A. Email: Chellasp@moffitt.usf.edu

Rb and Tumorigenesis, edited by Maurizio Fanciulli. ©2006 Eurekah.com and Springer Business+Science Media.

Brg1/Brm,[32] HP1,[33] SuV39H,[34] DNMT1,[35] as well as members of the polycomb family of proteins.[36,37] Inactivation of Rb by sequential phosphorylation events dissociates Rb from E2F as well as the corepressors, facilitating E2F-mediated transcription and cell cycle progression.[38] Not surprisingly, inactivation of Rb has been shown to be necessary for the progression of cells through the normal mammalian cell cycle.[3]

Multiple studies have shown that inactivation of Rb, either by phosphorylation, gene mutation, or binding of viral oncoproteins, could lead to uncontrolled cell proliferation resulting in tumorigenesis.[25,39,40] As has been reviewed elsewhere, components of the Rb cell cycle regulatory pathway are altered in almost all cancers; these include the mutation or inactivation of Rb, mutation of p16INK4 or other cdk inhibitors, or amplification or over-expression of cyclins.[10,41,42] Interestingly, while many molecules functioning upstream of Rb in the cell cycle regulatory cascade are altered in cancer, the major downstream targets of Rb, the E2F family members, are rarely perturbed. This is probably because of the dual functions of E2Fs, especially E2F1, in promoting cell proliferation as well as apoptosis.[17,28] Given that Rb can repress the transcriptional activity of E2F1, which is known to have proliferative and apoptotic functions, it is not surprising that inactivation of Rb is necessary for apoptosis induced by certain stimuli. This inactivation of Rb by apoptotic signals is also mediated by phosphorylation by kinases unrelated to the cyclin-cdk family.

Historically, Rb phosphorylation during cell cycle progression is thought to be mediated solely through the activation of cyclins and associated kinases.[43,44] D- and E-type cyclins, along with cdk4/6 or cdk2, sequentially phosphorylate Rb in multiple waves to effect complete inactivation.[38,45-47] Recent studies have suggested that other kinases, especially those in the MAP kinase cascade, can phosphorylate Rb as well, facilitating the phosphorylation by cyclin-cdks.[5,6] MAP kinases ERK1 and ERK2, as well as Raf-1 have been strongly implicated in the phosphorylation of Rb during cell cycle progression, prior to phosphorylation mediated by cyclins and cdks.[6,7] The first part of this review will examine the phosphorylation of Rb by noncyclin dependent kinases during cell cycle progression; the second part will deal with Rb phosphorylation during cellular apoptosis.

Regulation of Rb Phosphorylation during Cell Proliferation

MAP Kinase Mediated Phosphorylation of Rb

It had long been proposed that the ERK kinase could mediate the phosphorylation of Rb.[4,48,49] At the same time, it was not clear whether such events involved the activation of cyclins and cdks. A recent study shows that Rb is rapidly phosphorylated on Serine 795 upon treatment of vascular smooth muscle cells with angiotensin II or 5-hydroxytryptamine.[6] Though these agents function in pathways different than those utilized by classical growth factors, they are also able to initiate a mitogenic cascade. Ser 795 falls within the E2F1 binding domain of Rb, but mutating this site alone does not affect E2F1 binding.[50]

This study reports many novel aspects of Rb regulation, some of which might be signal or cell type specific. It was found that the rapid phosphorylation of Ser 795 was sensitive to PD98059, which is a MEK inhibitor that blocks the MAP kinase cascade. The kinetics of Rb phosphorylation is also interesting—it happens within 10 minutes of stimulation, after activation of the MAP kinase cascade but before the activation of cyclins and cdks. There was no phosphorylation of p107 or p130 at this time point, showing the specificity of the inactivation.[6]

One surprising aspect of the above study was that cyclin D and cdk4 were found to associate with MAP kinase upon stimulation with angiotensin II and 5-hydroxytryptamine. The authors propose that MAP kinase activation facilitates the interaction of cyclin D with cdk4.[6] The MEK inhibitor PD98059 prevented the interaction of cyclinD/cdk4 with MAP kinases. It was also found that stimulation of the cells with these agents led to the dissociation of E2F1 from Rb, an event that is not expected to happen upon phosphorylation of Ser 795. These findings raise the possibility that Rb could be regulated by MAP kinases in specific situations where hormone receptors activate an appropriate signaling cascade. The intriguing aspect, though, is that cyclin D and Cdk4 appear to be activated at very early stage in the cell cycle progression, quite different from a classic growth factor stimulation. Hence the relative role of MAP kinase per se in inactivating Rb remains unclear. Nevertheless, it can be concluded that MAP kinase cascade can inactivate Rb by mechanisms other than transcriptional induction of cyclins and activation of cdks, as is the case for growth factor signaling.

The role for MAP kinase cascade in inactivating Rb has been demonstrated indirectly in an elegant study.[51] It was found that while wild type mouse embryo fibroblast require the activation of the MAP kinase cascade to enter cell cycle, those lacking Rb did not. Interestingly, this event appeared to involve the entry of cell from G0 to G1 phase, not G1 to S phase; this suggests that inactivation of Rb by the MAP kinases plays a role in cell cycle progression facilitating the passage through the early stages, preceding the activation of cyclin dependent kinases.[51]

Involvement of MAP kinase cascade in regulating Rb function during the maturation of Xenopus oocytes has been proposed.[48] Maturation of Xenopus oocytes shows features of G1 as well as G2/M transition. In one study, microinjected Rb was found to be hypophosphorylated during prophase of oocytes; but it became hyperphosphorylated during meiotic maturation. The Rb phosphorylation observed during the maturation process was not dependent on cyclin D or cdk4; as expected, cyclin B/cdc2 was found to play a part in this inactivation. Interestingly, MAP kinase was also found to play a role in this inactivation. It was found that inhibition of MAP kinase cascade partially prevented the Rb phosphorylation. Similar studies showed that cyclin B/cdc2 as well as MAP kinase cascade play equally important roles in inactivating Rb during oocyte maturation.[48]

Overall, the sum of the above studies shows that components of the MAP kinase cascade can indeed inactivate Rb, in many cases directly, to facilitate cell cycle progression. These findings offer valuable insights into the mechanisms underlying cell proliferation and might be utilized to control proliferative disorders.

Regulation of Rb by Raf-1 Kinase

The Raf-1 kinase (c-Raf-1) is a vital link in the MAP kinase cascade and its activation is an essential feature of growth factor signaling.[52-54] One feature of the Raf-1 kinase is its narrow substrate specificity—only a limited number of molecules are known to be phosphorylated by Raf-1.[53] Surprisingly, Raf-1 was found to bind to Rb and p130 in yeast two-hybrid assays as well as in vitro binding assays.[5] There was no binding to the Rb family member, p107. The interaction seemed to be physiological, since immunoprecipitation-western blot experiments could detect a significant amount of Raf-1 bound to Rb and p130, but not p107, in asynchronous cells. The interaction was stringently regulated—there was no binding of Raf-1 to Rb in quiescent cells; but upon serum stimulation, the interaction could be detected in 30 minutes. It persisted for up to two hours before dissipating. Raf-1 is predominantly cytoplasmic and is activated at the membrane;[55] at the same time Rb is known to be a nuclear protein.[56] It was found that a small amount of Raf-1 translocated to the nucleus upon serum stimulation, and that binding to Rb occurred in the nucleus. Raf-1 kinase was able to phosphorylate Rb effectively in vitro; it appeared to be as good a substrate as MEK1.[5]

Raf-1 could functionally inactivate Rb and reverse Rb-mediated repression of E2F1 transcriptional activity as well as cell proliferation. These functions required a direct interaction between Raf-1 and Rb; deletion of the amino-terminal 28 amino acids of Raf-1 abolished its ability to bind Rb or reverse its function. It appears to be similar to viral oncoproteins in this aspect that a direct, stable binding is necessary to inactivate Rb; further, like viral oncoproteins, Raf-1 binds to the functional pocket domain of Rb. The similarity seems to end there, since viral oncoproteins dissociate E2F1 from Rb, Raf-1 binding does not. The kinase activity of Raf-1 was necessary for inactivation of Rb, since a kinase dead mutant could not inactivate Rb even after a direct binding.[5] Thus it appears that Raf-1 inactivates Rb after direct binding followed by phosphorylation.

The functional relevance of the Rb-Raf-1 interaction in cell cycle progression was studied by utilizing a peptide that could disrupt the binding of Raf-1 to Rb. Fine mapping studies of the Raf-1 protein showed that residues 10-18 in the amino-terminal region mediated the binding to Rb; 1 μM of this peptide could inhibit the Rb-Raf-1 interaction completely.[7] This was a specific competition—the peptide had no effect on the binding of other proteins to Rb or to Raf-1. Further, the binding of B-Raf to Rb was not affected by the Raf-1 peptide. Delivering the Raf-1 peptide conjugated to a carrier peptide, penetratin,[57] prevented the colocalization of Raf-1 with Rb in the nucleus. Additional biochemical experiments proved that the Raf-1 peptide could indeed disrupt the binding of Raf-1, but not E2F1, to Rb.

One pivotal aspect of the above study was the observation that disruption of the Rb-Raf-1 interaction leads to an inhibition of Rb phosphorylation, even after 16 hours of serum stimulation. Time course experiments suggested that the binding of Raf-1 to Rb precedes the binding of cyclin D; also, a low level of Rb phosphorylation could be observed 2 hours after serum stimulation, when there was no cyclin D bound to Rb. Raf-1 immunoprecipitated from cells stimulated with serum for 2 hours also could phosphorylate Rb; taken together, these experiments suggested that Raf-1 binding and phosphorylation of Rb is an event that precedes the phosphorylation of Rb by cyclin dependent kinases and probably facilitates the subsequent phosphorylation events. Inhibition of this priming phosphorylation event is preventing the subsequent steps, leaving Rb in a functional state. Interestingly, these results support the findings that cells lacking Raf-1 do not divide in culture.

Multiple lines of evidence suggest that Raf-1 mediated inactivation of Rb is a direct event. Earlier studies had shown that a mutant of Raf-1 that could not bind to Rb could not phosphorylate it. This implied that the MAP kinase cascade, which is the main mediator of Raf-1 function,[53] might not be involved in inactivating Rb. This possibility was examined by cotransfecting Raf-1 in the presence of dominant-negative MEK1 kinase or RKIP, which inhibits Raf-1 mediated activation of MEK1.[58-60] It was found that while RKIP as well as dominant-negative MEK1 inhibited Raf-1-mediated activation of AP1 transcription factor, which requires the MAP kinase cascade, the two molecules had no effect on Raf-1-mediated inactivation of Rb.

The fact that Raf-1 did not require the MAP kinase cascade to inactivate Rb raised some interesting possibilities. First, it revealed Rb to be a MAP-kinase independent target of Raf-1. Second, it also raised the possibility that cyclins and cdks may not be involved in Raf-1-mediated inactivation of Rb. This is based on the findings that the MAP kinase cascade induces the cyclin D promoter upon growth factor signaling, thus linking the signaling cascade with the cell cycle machinery. Since Raf-1 could bypass the MAP kinase cascade, it appeared possible that it might not be utilizing cyclins and cdks to inactivate Rb. Experiments were done to address this issue. It was found that over-expression of Raf-1 could overcome Rb function even when excess amounts of the cdk inhibitors, p16 or p21, were cotransfected; similarly, cotransfection of dominant-negative cdks had no effect on the Raf-1 mediated inactivation of Rb (Dasgupta et al, unpublished data). This suggests that an excess amount of Raf-1 can overcome Rb function even without the functioning of MAP kinases or cyclin-cdks. Though this is true of situations where Raf-1 is over-expressed, it is very likely that Raf-1 functions as a priming kinase during normal cell cycle progression—carrying out an initial phosphorylation, which facilitates the downstream phosphorylation events. Rb is known to be phosphorylated in multiple waves during cell cycle progression; the initial phosphorylation by Raf-1 appears to be an early, if not initiating, step in this cascade.

Studies from Doug Dean's lab had shown that such multiple steps of phosphorylation during cell cycle progression dissociates different proteins from Rb.[38] Thus E2F1 as well as HDAC1 required almost complete phosphorylation by cyclins D and E associated kinases. Since Raf-1 could not dissociate E2F1 from Rb, but could reverse Rb-mediated repression of E2F, the question was how Raf-1 de-represses E2F1. Chromatin immunoprecipitation assays as well as immunoprecipitation western blot assays showed that the binding of Raf-1 to Rb led to the dissociation of the chromatin remodeling protein Brg1 from Rb. Brg1 is known to be involved in Rb-mediated repression of E2F transcriptional activity, even when other corepressors are involved.[32,61] This result suggests that Raf-1, by dissociating Brg1 from Rb and the promoters repressed by Rb, facilitates the dissociation of other corepressors as well, rendering the promoter ready for transcriptional activation. Raf-1 seems to specifically dissociate Brg1 (and maybe Brm) from Rb; there was no change in the association of HDAC1 or HP1. This mode of transcriptional activation by Raf-1 is very different from the method used by viral oncoproteins like E1A and E7, which dissociate E2F1 from Rb, facilitating transcription.[25,39]

The effect of disrupting the Rb-Raf-1 interaction on cell proliferation was remarkable. It was found that delivering the peptide to serum-stimulated cells inhibited cells entering S-phase by about 50%.[7] This occurred in multiple cell lines, except, Saos-2, which lacked a functional Rb; this shows that the effects of the peptide was limited to cells having a functional Rb. Disrupting the Rb-Raf-1 interaction also could prevent cell proliferation induced by VEGF, which is known to activate Raf-1 kinase activity. Interestingly, disrupting the Rb-Raf-1 interaction inhibited VEGF-induced angiogenic tubule formation of HAECs in matrigel significantly; this was accompanied by a disruption of the Rb-Raf-1 interaction, but there was no change in the levels of MAP kinase activity in the cells. The inhibition of angiogenic tubule formation was due to the inhibitory effects of the peptide on the adhesion, migration and invasion of HAECs, processes which are necessary for angiogenic tubule formation. These experiments suggest that the Rb-Raf-1 interaction contributes to various processes like cell proliferation and angiogenesis, which facilitates oncogenesis and tumor progression. Given these results, it was examined whether Rb-Raf-1 interaction is altered in human tumors compared to normal tissue. Whole-cell lysates were prepared from ten nonsmall cell lung carcinomas resected from patients and the amount of Rb-Raf-1 interaction assessed by immunoprecipitation-western blotting. It was found that in eight out of the ten cases, there was more Rb-Raf-1 interaction in the tumors compared to the adjacent normal tissues (Dasgupta et al, unpublished results). This raises the possibility that the elevated Rb-Raf-1 interaction has contributed to the oncogenic process. This is not surprising, given the stimulatory role this interaction plays in cell proliferation and angiogenesis.

Since agents that can inhibit cell proliferation and angiogenesis have the potential to be good anti-cancer agents, the ability of the Raf-1 peptide to inhibit tumor growth in vivo was examined. It was found that administration of the peptide conjugate intratumorally to nonsmall cell carcinomas xenotransplanted into nude mice led to a reduction in the tumor volume by about 79%. In addition, the tumors had around 50% less neovasculature compared to untreated tumor, as seen by CD31 staining. These experiments suggest that agents capable of disrupting the Rb-Raf-1 interaction have the potential to be of value as anti-cancer agents.

Thus the Rb-Raf-1 interaction appears to be a vital event that occurs during the early part of mitogenesis and seems to be necessary to facilitate cell cycle progression. Its ability to facilitate the inactivation of Rb makes it a good target for developing agents to combat cancer.

Regulation of Rb during Apoptosis

It is well-established that Rb has potent anti-apoptotic activities and that many signals have to overcome Rb function to induce apoptosis.[3,8] The question as to how Rb is inactivated upon apoptotic signaling has been addressed in the recent years.[11-13] These studies show that Rb inactivation during cellular apoptosis is mainly independent of cyclins and cdks. It appears that kinases like p38, ASK1 and in some cases, JNK1, might play a role in the apoptotic process. Further analysis of the regulation of Rb function during apoptosis might eventually enable us to modulate these pathways to selectively kill cancer, but not normal, cells.

Regulation of Rb by p38 Kinase

Rb protein is known to have potent anti-proliferative properties as well, and many signals have to overcome Rb function to induce apoptosis. The MAP kinase family member p38 is activated by many apoptotic signals and inhibition of the p38 kinase activity can partially inhibit apoptosis induced by agents like TNFα or Fas.[62,63] Similarly, the stress activated kinase JNK1 is also known to be activated upon apoptotic signaling in certain cell lines. Since the cell cycle machinery contributes to the apoptotic process, these kinases seemed to be likely candidates to affect Rb function upon apoptotic signaling. Inactivation of Rb by the p38 kinase has been described in two systems: first in the T-cell lymphoma cell line Jurkat[15] and second in a neuronal system.[12] Similar findings were also reported in endothelial cells.[64] Indeed, over-expression of p38 kinase was found to induce cell proliferation as seen by a colony formation assay in Jurkat cells; at the same time, JNK1 was found to inhibit it.[11] Cotransfection experiments showed that JNK1 could repress E2F1-mediated transcriptional activity while p38 could overcome Rb-mediated repression of E2F1; thus the two kinases seemed to have opposite effects on cell proliferation as well as the Rb-E2F pathway.[11]

As in the case of the Raf-1 kinase, p38 could also reverse Rb mediated repression of E2F1 in a phosphorylation dependent manner—a kinase deficient mutant of p38 kinase had no effect on Rb function.[11] Induction of p38 kinase by treatment of cells with TNFα or an anti-Fas antibody also resulted in reversal of transcriptional repression, in a p38 dependent manner. While JNK1 was found to phosphorylate E2F1 and directly prevent its binding to DNA, it did not appear to have any direct effects on Rb function. On the other hand, cotransfection or induction of p38 kinase in Jurkat T-cell lymphoma cells led to an increase in Rb phosphorylation. This coincided with the dissociation of E2F1 from Rb, facilitating transcriptional activation. Stimulation of Jurkat cells with an anti-Fas antibody led to Rb phosphorylation as well as transcriptional activation of E2F1; this could not be inhibited by olomoucine or roscovitine, two inhibitors of cyclin dependent kinases. This indicated that Fas stimulation of cells might be inactivating Rb in a cyclin-cdk independent manner.

Rb could be phosphorylated by p38 kinase effectively in vitro; it was very similar to Raf-1 in this aspect.[11] This phosphorylation was sensitive to SB203580, a p38 kinase inhibitor. Phosphorylation of Rb in response to Fas stimulation also was sensitive to SB203580, suggesting that Fas is inactivating Rb predominantly by a p38 kinase dependent manner. Rb phosphorylation in response to Fas seemed to follow fairly fast kinetics, in that Rb was phosphorylated within 30 minutes; this persisted for up to 2 hours; Rb was proteolytically cleaved within 6 hours. Studies from other labs had shown that Rb is cleaved in a caspase-dependent manner during Fas-induced apoptosis;[9,65,66] the phosphorylation of Rb by p38 appears to precede this degradation event. It is not yet clear whether the p38-mediated phosphorylation is necessary for Rb to be cleaved by the caspases.

The role of cyclin-cdks in p38-mediated inactivation of Rb was assessed. It was found that over-expression of dominant-negative cdks did not affect Fas-mediated phosphorylation and inactivation of Rb; at the same time, a dominant-negative p38 kinase completely inhibited the inactivation. Same results were obtained when p38 kinase was over-expressed.[11] These results suggest that like Raf-1 kinase, p38 kinase can also inactivate Rb independent of cyclins and cdks. On the contrary, there are two major differences between Raf-1 and p38 mediated inactivation of Rb. While Raf-1 was found to physically interact with Rb and this was necessary for the inactivation, IP-western blots could not detect a stable interaction of p38 kinase with Rb. Further, Raf-1 does not dissociate E2F1 from Rb, unlike p38 kinase: rather, it dissociates Brg1, relieving the repression of transcription. Nevertheless, it appears that multiple kinases outside the cyclin-cdk family can inactivate Rb independently, and relieve its repression of E2F by different mechanisms.

The independence of p38-mediated inactivation of Rb from cyclin dependent kinases raised the questions whether p38 acts upstream of cyclin-cdks, as in the case of Raf-1, or functions in totally different settings. It was found that serum-stimulation of quiescent cells in the presence of SB203580 could inactivate Rb efficiently, enhancing E2F-mediated transcription.[13] This suggested that p38 kinase may not be playing a role in inactivating Rb during cell cycle progression. It was also found that phosphorylation of Rb upon serum stimulation is not affected by SB203580, confirming the above result. Similarly, the dissociation of E2F1 from Rb occurred efficiently during serum stimulation, even when SB203580 was present. These are events that were totally blocked by the p38 kinase inhibitor during Fas stimulation.[13] These results suggest that p38 kinase functions to inactivate Rb only in apoptotic settings, and does not play a role in modulating cell cycle progression.

The functional dichotomy of p38 during cell cycle progression and apoptosis was extended to the two other Rb family members. It was found that p38 kinase could inactivate p107 fairly efficiently. At the same time, it had no effect on the p130 protein.[13] This is in stark contrast to Raf-1, which could inactivate Rb and p130, but not p107.[5] Similar effects were noticed when Fas stimulation was used to induce p38, rather than over-expression. It is intriguing that two different Rb family members are affected differentially during apoptosis and cell cycle progression by different non cyclin dependent kinases: it appears that activation of the cyclin-cdk pathway would be essential to inactivate all the Rb family members completely. It is also possible that inactivation of Rb and p107 alone is sufficient to facilitate the apoptosis initiated by Fas signaling.

The finding that p38 and cdks phosphorylate and inactivate Rb during different physiological processes and the fact that they specifically target specific Rb family members raised the possibility that these kinases are phosphorylating different sites on the Rb protein. Jean Wang's lab had created various phosphorylation site mutants (PSM mutants) of the Rb protein that are resistant to phosphorylation by cyclin-cdks.[67] The ability of such a mutant to respond to Fas as well as p38 signaling was examined. It was found that cotransfection of cyclin D did not affect the phosphorylation of a PSM-7 mutant of Rb (one with seven phosphorylation sites mutated). Interestingly, this mutant could be phosphorylated efficiently by over-expressing the p38 kinase. This suggested that cyclins and p38 kinase are phosphorylating Rb on different sites. Similar results were obtained when cyclin-cdks were activated by serum stimulation—while they could inactivate wild-type Rb, as expected, there was no effect on the PSM-7 Rb. Fas stimulation, on the other hand, led to the inactivation of both wild-type as well as PSM-7 Rb. It can be concluded that apoptotic signals like Fas stimulation can inactivate Rb utilizing the p38 kinase, which phosphorylates Rb on sites distinct from cdk phosphorylation sites. This is another interesting example of a specific signal affecting components of the cell cycle machinery utilizing non cyclin-dependent kinases.

Similar to the Jurkat system, p38 kinase has been found to induce Rb inactivation upon Fas stimulation of cerebellar granule neurons.[12] It was found that Fas is expressed on the surface of these neurons and its activation by a ligand or an antibody led to neuronal apoptosis. This was found involve the Rb-E2F pathway. Rb inactivation upon Fas stimulation of the cerebellar granule neurons led to the phosphorylation of Rb; this was again, dependent on the p38 kinase. Inhibition of the p38 kinase, but not the cyclin-cdks by a variety of inhibitors prevented Rb phosphorylation.[12] Similar to Jurkat cells, E2F1 was found to be dissociated from Rb upon Fas stimulation. The authors speculate that activation of E2F-mediated transcriptional activity, via the p38-mediated inactivation of Rb contributes to Fas-induced apoptosis of cerebellar granule neurons. This possibility was examined by an elegant experiment. Cerebellar granule neurons from mice null for E2F1 were stimulated with an anti-Fas antibody and neuronal apoptosis measured. It was found that neurons from the E2F1 null mice were less susceptible to the apoptotic effects of Fas; this supports the contention that inactivation of Rb by the p38 kinase leads to the activation of E2F1-mediated transcription, leading to cellular apoptosis. The authors propose two mechanisms to explain this: inactivation of Rb and activation of E2F1 might result either in the activation of pro-apoptosis genes, or the suppression of survival genes. This is a very reasonable hypothesis, given the fact that many pro-apoptotic genes (p73, Apaf-1, caspase 3, caspase 7, etc) are directly upregulated by E2F1, while certain other genes like p53 are indirectly induced by E2F1, through the involvement of p16ARF and mdm2.

Similar studies on the role of JNK1 and p38 on Rb and E2F1 were carried out on endothelial cells.[64] It was found that JNK1 could bind to and phosphorylate E2F1, inhibiting its transcriptional activity. Though the basic finding that JNK1 can phosphorylate and inactivate E2F1 is very similar to the one made on Jurkat cells upon TNFα treatment, the mechanism proposed is different. Whereas JNK1 was found to bind to DP1 and phosphorylate E2F1 in Jurkat cells, the binding appeared to be specific for E2F1 in this study.[11] The basis for this discrepancy is unknown. At the same time, this study also confirms earlier findings on p38 kinase in Jurkat cells. The authors show that over-expression of a constitutively active p38 kinase could overcome the repression of E2F1 mediated by TNFα by facilitating its dissociation from Rb. These results also show that while TNFα can induce both JNK1 and p38 kinases in the cells, the extent of activation might vary depending on the cell type. This could lead to either the activation or repression of E2F-mediated transcription.

Regulation of Rb Function by JNK1

One study that obtained different results on Rb inactivation was conducted in multiple myeloma cells.[14] It was found that γ-irradiation leads to the activation of JNK1, which bound to Rb. This study also showed that JNK1 could bind to Rb in vitro and in vivo, a result not reported by other groups. The major difference was in the technique used for detecting the interaction in vivo: this group used far-western blotting to show the association of JNK1 with Rb, while other groups had not. The in vitro results are more difficult to explain, since many of the experiments done in other labs could not find the binding of JNK1 to Rb. The study in question also shows that activated JNK1 could phosphorylate Rb at a carboxy terminal site and that deletion of this domain prevents the phosphorylation by JNK1.[14] While these results need to be reproduced in other cell lines and apoptotic signals, this appears to be another example by which Rb is regulated by a kinase unrelated to the cyclin-cdk family.

The final conclusion that can be drawn from these studies is that apoptotic signals can contact the Rb protein via the p38 kinase, independent of any contribution from cyclin-cdks, leading to the activation of E2F function resulting in apoptosis.

Regulation of Rb Function by Apoptosis Signal Regulated Kinase 1 (ASK1)

ASK1 is a MAP-kinase-kinase-kinase like Raf-1, but functions in the apoptotic signaling process.[68,69] It is known to be activated by various apoptotic stimuli, including TNFα and Fas.[70,71] ASK1 can effectively activate p38 kinase, and in certain cases JNK1. It is known to be maintained in an inactive complex with other cellular proteins like thioredoxin, HSP72 as well as 14-3-3 family members.[68,69,72-74] It has been shown that the Raf-1 kinase could physically interact with ASK1 and inhibit its apoptotic properties, independent of the MAP kinase cascade.[75]

An examination of ASK1 sequence revealed the presence of an LXCXE motif that is used by viral oncoproteins like adenovirus E1A, SV40 large T-antigen and HPV E7 to bind to Rb.[40,76-78] Given that another MAP-kinase-kinase-kinase, Raf-1, could bind to Rb, the ability of ASK1 to do the same was examined. It was found that ASK1 could bind to Rb, p107 and p130 in vitro; the binding required a functional pocket domain of Rb.[15] Mutating the LXCXE domain of ASK1 abolished its binding to Rb suggesting that it bound to Rb using the same motif as viral oncoproteins. This was further confirmed by the finding that adenovirus E1A protein could compete the binding of ASK1 to Rb. In addition to this, Raf-1 kinase was also efficient in displacing ASK1 from Rb in vitro as well as in vivo; this suggests that a different MAP-kinase-kinase-kinase could be binding to Rb upon mitogenic or apoptotic signaling. Raf-1 has been shown to bind to Rb as well as ASK1; it was found that it was the binding of Raf-1 to ASK1 that prevented the latter from binding to Rb, since a mutant Raf-1 incapable of binding to Rb could also prevent the ASK1-Rb interaction.[15]

Many apoptotic signals could induce the binding of ASK1 to Rb in vivo, as seen by IP-western blot analysis. Thus TNFα, Fas and H_2O_2 could induce this interaction within 30 minutes in multiple cell lines, including Jurkat, Ramos, 3T3 and Human Aortic Endothelial Cells (HAEC). The interaction appeared at 30 minutes after exposure to the apoptotic signals and vanished within 2-4 hours, depending on the cell line. The anti-oxidant N-acetyl cysteine could inhibit the association of ASK1 with Rb, showing that activation of ASK1 by oxidative stress is necessary for it to bind to Rb. It was also seen that while serum stimulation of quiescent Ramos cells induced the binding of Raf-1 to Rb, there was no binding of ASK1; similarly, treatment of Ramos cells with TNFα led to the binding of ASK1, not Raf-1, to Rb. This confirmed the possibility that different MAP-kinase-kinase-kinases associate with Rb and inactivate it, depending on the signal.

An immunofluorescence experiment showed that ASK1 does translocate to the nucleus upon stimulation with apoptotic agents like TNFα; further, a double immunofluorescence experiment showed that such signals not only facilitate the nuclear translocation of ASK1, but also induces its association with Rb. Like Raf-1, over-expression of ASK1 was found to reverse Rb-mediated repression of E2F1 transcriptional activity.[15] This appeared to be due to a reduction in the amount of E2F1 associated with Rb after the binding of ASK1. Thus ASK1 appeared to affect E2F1 activity in a manner similar to viral oncoproteins as well as E1A. Since kinases that dissociate E2F1 from Rb can phosphorylate Rb directly, the ability of ASK1 to phosphorylate Rb was tested. It was found that ASK1 could phosphorylate Rb efficiently in vitro; this required a direct binding to Rb. This conclusion was reached based on the finding that an ASK1 mutant that could not bind to Rb could phosphorylate myelin basic protein, but not Rb. Thus like Raf-1, an extra-cellular signal leads the binding of ASK1 to Rb, leading to the phosphorylation and inactivation of the latter.

Over-expression of ASK1 is known to induce apoptosis, mainly in cell lines like HeLa and 293s, which harbor viral oncoproteins.[68] Since these viral oncorproteins inactivate Rb, and since Rb is known to have potent anti-apoptotic properties, the possibility that ASK1 has to inactivate Rb to induce apoptosis was examined. It was found that over-expression of Rb could inhibit ASK1-mediated induction of apoptosis in 3T3 cells; at the same time, an Rb pocket domain mutant that could not bind to ASK1 could not inhibit ASK1-induced apoptosis. Further, it was observed that an ASK1 LXCXE domain mutant that could not bind to Rb could not induce apoptosis either. These experiments collectively indicate that ASK1 has to overcome the anti-apoptotic activity of Rb to induce apoptosis.

Since E2F1 is known to regulate the expression of mitogenic as well as pro-apoptotic proteins and since ASK1 was found to relieve Rb-mediated repression of E2F1, it was examined how ASK1 affects the expression of E2F target genes. It was found that the expression of the p53 family member, p73,[79-82] was enhanced in cells over-expressing ASK1; p73 is known to be a direct transcriptional target of E2F1.[83] This was found to be true for 3T3 as well as Ramos cells. In all the cases, the enhancement of p73 occurred only in cells transfected with a wild type ASK1 construct; those transfected with an Rb-binding mutant of ASK1 had no effect on p73 expression, suggesting that the binding of ASK1 to Rb contributes to the induction of the p73 gene.

Studies aimed at understanding how ASK1 stimulation leads to the induction of a specific pro-apoptotic, but not mitogenic promoter, led to some interesting findings. Chromatin immunoprecipitation (ChIP) assays were utilized to examine the binding of E2F1 as well as Rb to pro-apoptotic as well as mitogenic promoters like Cdc25A. It was found that stimulation of Ramos cells with TNFα led to the increased binding of E2F1 to the p73P1 promoter in vivo. At the same time, the amount of Rb associated with this promoter was reduced significantly. On the contrary, E2F3, which is known to have mainly proliferative effects[84,85] was dissociated from the p73 promoter upon stimulation with TNFα. ChIP assays were also done on the Cdc25A promoter to see whether TNFα stimulation affected the occupancy of this promoter. It was found that TNFα stimulation led to the reduced binding of E2F1, while enhancing the

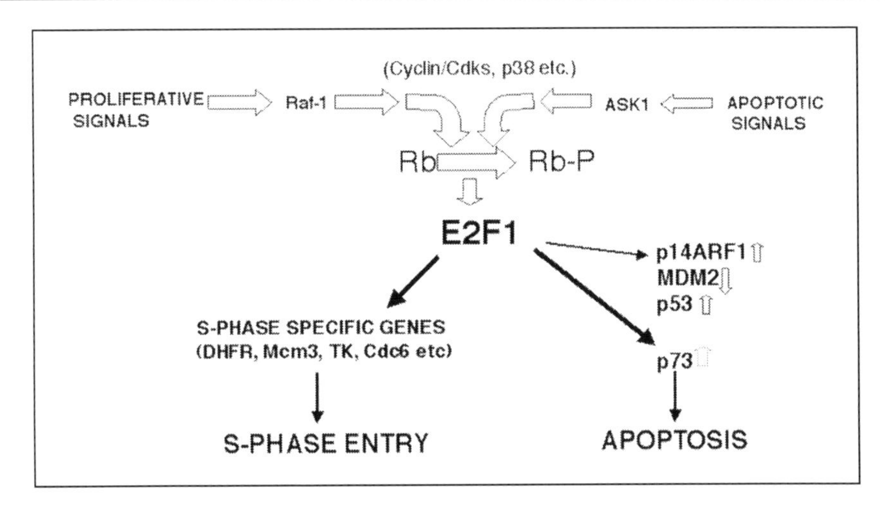

Figure 1. Inactivation of Rb by proliferative signals involves phosphorylation events mediated by Raf-1 and may be MAP kinases, which facilitate further phosphorylation by cyclin/cdks. This will lead to enhancement of E2F1 activity, leading to the expression of proliferative promoters. On the other hand, apoptotic signals like TNFα and Fas can inactivate Rb through ASK1 and p38 kinases, independent of cyclin/cdks. Activation of E2F1 in this situation will lead to the expression of pro-apoptotic genes, resulting in cell death.

amount of Rb bound to this promoter significantly. The amount of E2F3 bound remained constant. It thus appears that stimulation with TNFα leads to dissociation of Rb from pro-apoptotic promoters, facilitating their expression; concurrently, there is increased binding of Rb to proliferative promoters, leading to their repression. This gives the impression that Rb is functioning as a switch facilitating apoptosis or proliferation, depending on the signal a cell receives. This might partially explain why abrogation of the Rb regulatory pathways contribute to the onset of almost all cancer types.

Conclusions

The above findings throw light on a relatively unknown facet of Rb function- that it can be regulated by kinases outside the spectrum of cyclins and cdks. Further, the regulation of Rb function in response to apoptotic signaling almost invariably occurs without any mediation of cyclins and cdks. The same cannot be said of mitogenic signaling—while it is clear that kinases belonging to the MAP kinase cascade can inactivate Rb, it appears that this facilitates the subsequent inactivation steps known to be mediated by cyclins and cdks. As shown in Figure 1, induction of Raf-1/MAP kinase cascade initiates the phosphorylation events on Rb in response to mitogenic signaling. This will lead to complete phosphorylation mediated by cyclin/cdks, resulting in the activation of the appropriate E2Fs and induction of promoters necessary for cell cycle progression. On the other hand, inactivation of Rb by ASK1 or p38 kinases in response to apoptotic signals will lead to cyclin-independent phosphorylation of Rb and activation of genes like p73 or p53, leading to apoptosis. The identification of these novel regulatory modes for Rb, its regulation by Raf-1, for example, can be expected to open new avenues for the development of novel agents for cancer therapy.

Acknowledgements

The authors wish to thank members of the Chellappan lab for helpful comments and suggestions. Studies in the authors laboratory are funded by grants CA63136 and CA77301 from the NIH. Apologies to many authors whose work could not be cited directly.

References

1. Weinberg RA. The retinoblastoma protein and cell cycle control. Cell 1995; 81:323-330.
2. Sherr CJ. The ins and outs of RB: Coupling gene expression to the cell cycle clock. Trends Cell Biol 1994; 4:15-18.
3. Harbour JW, Dean DC. Rb function in cell-cycle regulation and apoptosis. Nat Cell Biol 2000; 2:E65-67.
4. Taya Y. RB kinases and RB-binding proteins: New points of view. Trends Biochem Sci 1997; 22:14-17, [Review] [50 refs].
5. Wang S, Ghosh RN, Chellappan SP. Raf-1 physically interacts with Rb and regulates its function: A link between mitogenic signaling and cell cycle regulation. Mol Cell Biol 1998; 18:7487-7498.
6. Garnovskaya MN, Mukhin YV, Vlasova TM et al. Mitogen-induced rapid phosphorylation of serine 795 of the retinoblastoma gene product in vascular smooth muscle cells involves ERK activation. J Biol Chem 2004; 279:24899-24905.
7. Dasgupta P, Sun J, Wang S et al. Disruption of the Rb-Raf-1 interaction Inhibits Tumor growth and Angiogenesis. Molecular Cellular Biology 2004; 24:9527-9541.
8. Chau BN, Wang JY. Coordinated regulation of life and death by RB. Nat Rev Cancer 2003; 3:130-138.
9. Tan X, Wang JY. The caspase-RB connection in cell death. Trends Cell Biol 1998; 8:116-120.
10. Sherr CJ, McCormick F. The RB and p53 pathways in cancer. Cancer Cell 2002; 2:103-112.
11. Wang S, Nath N, Minden A et al. Regulation of Rb and E2F by signal transduction cascades: Divergent effects of JNK1 and p38 kinases. Embo J 1999; 18:1559-1570.
12. Hou ST, Xie X, Baggley A et al. Activation of the Rb/E2F1 pathway by the nonproliferative p38 MAPK during Fas (APO1/CD95)-mediated neuronal apoptosis. J Biol Chem 2002; 277:48764-48770.
13. Nath N, Wang S, Betts V et al. Apoptotic and mitogenic stimuli inactivate Rb by differential utilization of p38 and cyclin-dependent kinases. Oncogene 2003; 22:5986-5994.
14. Chauhan D, Hideshima T, Treon S et al. Functional interaction between retinoblastoma protein and stress-activated protein kinase in multiple myeloma cells. Cancer Res 1999; 59:1192-1195.
15. Dasgupta P, Betts V, Rastogi S et al. Direct binding of apoptosis signal-regulating kinase 1 to retinoblastoma protein: Novel links between apoptotic signaling and cell cycle machinery. J Biol Chem 2004; 279:38762-38769.
16. Paggi MG, Baldi A, Bonetto F et al. Retinoblastoma protein family in cell cycle and cancer: A Review. J Cell Biochem 1996; 62:418-430.
17. Johnson DG, Schneider-Broussard R. Role of E2F in cell cycle control and cancer. Front Biosci 1998; 3:d447-448.
18. Ewen ME. The cell cycle and the retinoblastoma protein family. Cancer & Metastasis Reviews 1994; 13:45-66.
19. Ewen ME. Regulation of the cell cycle by the Rb tumor suppressor family. Results Probl Cell Differ 1998; 22:149-179.
20. Hatakeyama M, Weinberg RA. The role of RB in cell cycle control. Prog Cell Cycle Res 1995; 1:9-19.
21. Chellappan SP, Hiebert S, Mudryj M et al. The E2F transcription factor is a cellular target for the RB protein. Cell 1991; 65:1053-1061.
22. Hiebert SW, Chellappan SP, Horowitz JM et al. The interaction of RB with E2F coincides with an inhibition of the transcriptional activity of E2F. Genes Dev 1992; 6:177-185.
23. Morris EJ, Dyson NJ. Retinoblastoma protein partners. Adv Cancer Res 2001; 82:1-54.
24. Harbour JW, Dean DC. The Rb/E2F pathway: Expanding roles and emerging paradigms [In Process Citation]. Genes Dev 2000; 14:2393-2409.
25. Nevins JR. E2F: A link between the Rb tumor suppressor protein and viral oncoproteins. Science 1992; 258:424-429.
26. Stevaux O, Dyson NJ. A revised picture of the E2F transcriptional network and RB function. Curr Opin Cell Biol 2002; 14:684-691.
27. Trimarchi JM, Lees JA. Transcriptionsibling rivalry in the E2F family. Nat Rev Mol Cell Biol 2002; 3:11-20.
28. Ginsberg D. E2F1 pathways to apoptosis. FEBS Lett 2002; 529:122-125.
29. DeGregori J. The genetics of the E2F family of transcription factors: Shared functions and unique roles. Biochim Biophys Acta 2002; 1602:131-150.
30. Luo RX, Postigo AA, Dean DC. Rb interacts with histone deacetylase to repress transcription. Cell 1998; 92:463-473.
31. Ferreira R, Naguibneva I, Pritchard LL et al. The Rb/chromatin connection and epigenetic control: Opinion. Oncogene 2001; 20:3128-3133.

32. Dunaief JL, Strober BE, Guha S et al. The retinoblastoma protein and BRG1 form a complex and cooperate to induce cell cycle arrest. Cell 1994; 79:119-130.
33. Nielsen SJ, Schneider R, Bauer UM et al. Rb targets histone H3 methylation and HP1 to promoters. Nature 2001; 412:561-565.
34. Vaute O, Nicolas E, Vandel L et al. Functional and physical interaction between the histone methyl transferase Suv39H1 and histone deacetylases. Nucleic Acids Res 2002; 30:475-481.
35. Robertson KD, Ait-Si-Ali S, Yokochi T et al. DNMT1 forms a complex with Rb, E2F1 and HDAC1 and represses transcription from E2F-responsive promoters. Nat Genet 2000; 25:338-342.
36. Dahiya A, Wong S, Gonzalo S et al. Linking the Rb and polycomb pathways. Mol Cell 2001; 8:557-569.
37. Harbour JW, Dean DC. Chromatin remodeling and Rb activity. Curr Opin Cell Biol 2000; 12:685-689.
38. Harbour JW, Luo RX, Dei Santi A et al. Cdk phosphorylation triggers sequential intramolecular interactions that progressiively block Rb functions as cells move through G1. Cell 1999; 98:859-869.
39. Chellappan S, Kraus VB, Kroger B et al. Adenovirus E1A, simian virus 40 tumor antigen, and human papillomavirus E7 protein share the capacity to disrupt the interaction between transcription factor E2F and the retinoblastoma gene product. Proc Natl Acad Sci USA 1992; 89:4549-4553.
40. Nevins JR. The Rb/E2F pathway and cancer. Hum Mol Genet 2001; 10:699-703.
41. Kamb A. Cell cycle regulators and cancer. Trends Genet 1995; 11:136-140.
42. Kamb A, Gruis NA, Weaver-Feldhaus J et al. A cell cycle regulator potentially involved in genesis of many tumor types. Science 1994; 264:436-440.
43. Kato J, Matsushime H, Hiebert SW et al. Direct binding of cyclin D to the retinoblastoma gene product (pRb) and pRb phosphorylation by the cyclin D-dependent kinase CDK4. Genes Dev 1993; 7:331-342.
44. Reed SI. Control of the G1/S transition. Cancer Surv 1997; 29:7-23.
45. Lundberg AS, Weinberg RA. Functional inactivation of the retinoblastoma protein requires sequential modification by at least two distinct cyclin-cdk complexes. Mol Cell Biol 1998; 18:753-761.
46. Mittnacht S, Lees JA, Desai D et al. Distinct subpopulations of retinoblastoma proteins show a distinct pattern of phosphorylation. EMBO J 1994; 13:118-127.
47. Hinds PW, Mittnacht S, Dulic V et al. Regulation of retinoblastoma protein functions by ectopic expression of human cyclins. Cell 1992; 70:993-1006.
48. Taieb F, Karaiskou A, Rime H et al. Human retinoblastoma protein (Rb) is phosphorylated by cdc2 kinase and MAP kinase in Xenopus maturing oocytes. FEBS Lett 1998; 425:465-471.
49. Mittnacht S, Paterson H, Olson MF et al. Ras signalling is required for inactivation of the tumour suppressor pRb cell-cycle control protein. Curr Biol 1997; 7:219-221.
50. Knudsen ES, Wang JY. Dual mechanisms for the inhibition of E2F binding to RB by cyclin-dependent kinase-mediated RB phosphorylation. Mol Cell Biol 1997; 17:5771-5783.
51. D'Abaco GM, Hooper S, Paterson H et al. Loss of Rb overrides the requirement for ERK activity for cell proliferation. J Cell Sci 2002; 115:4607-4616.
52. Daum G, Eisenmann-Tappe I, Fries HW et al. The ins and outs of Raf kinases. Trends Biochem Sci 1994; 19:474-480.
53. Marshall CJ. Cell signalling. Raf gets it together. Nature 1996; 383:127-128.
54. Dhillon AS, Kolch W. Untying the regulation of the Raf-1 kinase. Arch Biochem Biophys 2002; 404:3-9.
55. Marais R, Light Y, Paterson HF et al. Ras recruits Raf-1 to the plasma membrane for activation by tyrosine phosphorylation. Embo J 1995; 14:3136-3145.
56. Mancini MA, Shan B, Nickerson JA et al. The retinoblastoma gene product is a cell cycle-dependent, nuclear matrix-associated protein. Proceedings of the National Academy of Sciences of the United States of America 1994; 91:418-422.
57. Derossi D, Chassaing G, Prochiantz A. Trojan peptides: The penetratin system for intracellular delivery. Trends Cell Biol 1998; 8:84-87.
58. Yeung K, Seitz T, Li S et al. Suppression of Raf-1 kinase activity and MAP kinase signalling by RKIP. Nature 1999; 401:173-177.
59. Yeung KC, Rose DW, Dhillon AS et al. Raf kinase inhibitor protein interacts with NF-kappaB-inducing kinase and TAK1 and inhibits NF-kappaB activation. Mol Cell Biol 2001; 21:7207-7217.
60. Yeung K, Janosch P, McFerran B et al. Mechanism of suppression of the Raf/MEK/extracellular signal-regulated kinase pathway by the raf kinase inhibitor protein. Mol Cell Biol 2000; 20:3079-3085.
61. Dahiya A, Gavin MR, Luo RX et al. Role of the LXCXE binding site in Rb function. Mol Cell Biol 2000; 6799-6805.
62. Paul A, Wilson S, Belham CM et al. Stress-activated protein kinases: Activation, regulation and function. Cell Signal 1997; 9:403-410.

63. Kyriakis JM, Avruch J. Protein kinase cascades activated by stress and inflammatory cytokines. Bioessays 1996; 18:567-577.
64. Kishore R, Luedemann C, Bord E et al. Tumor necrosis factor-mediated E2F1 suppression in endothelial cells: Differential requirement of c-Jun N-terminal kinase and p38 mitogen-activated protein kinase signal transduction pathways. Circ Res 2003; 93:932-940.
65. Dou QP, An B, Antoku K et al. Fas stimulation induces RB dephosphorylation and proteolysis that is blocked by inhibitors of the ICE protease family. J Cell Biochem 1997; 64:586-594.
66. Tan X, Martin SJ, Green DR et al. Degradation of retinoblastoma protein in tumor necrosis factor- and CD95-induced cell death. J Biol Chem 1997; 272:9613-9616.
67. Knudsen ES, Wang JY. Differential regulation of retinoblastoma protein function by specific Cdk phosphorylation sites. J Biol Chem 1996; 271:8313-8320.
68. Ichijo H, Nishida E, Irie K et al. Induction of apoptosis by ASK1, a mammalian MAPKKK that activates SAPK/JNK and p38 signaling pathways. Science 1997; 275:90-94.
69. Matsuzawa A, Ichijo H. Molecular mechanisms of the decision between life and death: Regulation of apoptosis by apoptosis signal-regulating kinase 1. J Biochem (Tokyo) 2001; 130:1-8.
70. Chang HY, Nishitoh H, Yang X et al. Activation of apoptosis signal-regulating kinase 1 (ASK1) by the adapter protein Daxx. Science 1998; 281:1860-1863.
71. Chen Z, Seimiya H, Naito M et al. ASK1 mediates apoptotic cell death induced by genotoxic stress. Oncogene 1999; 18:173-180.
72. Ichijo H. From receptors to stress-activated MAP kinases. Oncogene 1999; 18:6087-6093.
73. Takeda K, Hatai T, Hamazaki TS et al. Apoptosis signal-regulating kinase 1 (ASK1) induces neuronal differentiation and survival of PC12 cells. J Biol Chem 2000; 275:9805-9813.
74. Tobiume K, Matsuzawa A, Takahashi T et al. ASK1 is required for sustained activations of JNK/p38 MAP kinases and apoptosis. EMBO Rep 2001; 2:222-228.
75. Chen J, Fujii K, Zhang L et al. Raf-1 promotes cell survival by antagonizing apoptosis signal-regulating kinase 1 through a MEK-ERK independent mechanism. Proc Natl Acad Sci USA 2001; 98:7783-7788.
76. Moran E. DNA tumor virus transforming proteins and the cell cycle. Curr Opin Genet Dev 1993; 3:63-70.
77. Nevins JR. Disruption of cell-cycle control by viral oncoproteins. Biochemical Society Transactions 1993; 21:935-938, [Review].
78. Nevins JR. Cell cycle targets of the DNA tumor viruses. Current Opinion in Genetics & Development 1994; 4:130-134.
79. Levrero M, De Laurenzi V, Costanzo A et al. Structure, function and regulation of p63 and p73. Cell Death Differ 1999; 6:1146-1153.
80. Ichimiya S, Nakagawara A, Sakuma Y et al. p73: Structure and function. Pathol Int 2000; 50:589-593.
81. Davis PK, Dowdy SF. p73. Int J Biochem Cell Biol 2001; 33:935-939.
82. Stiewe T, Putzer BM. p73 in apoptosis. Apoptosis 2001; 6:447-452.
83. Stiewe T, Putzer BM. Role of the p53-homologue p73 in E2F1-induced apoptosis. Nat Genet 2000; 26:464-469.
84. Leone G, DeGregori J, Yan Z et al. E2F3 activity is regulated during the cell cycle and is required for the induction of S phase. Genes Dev 1998; 12:2120-2130.
85. Wu L, Timmers C, Maiti B et al. The E2F1-3 transcription factors are essential for cellular proliferation. Nature 2001; 414:457-462.

Chapter 6

Diverse Regulatory Functions of the E2F Family of Transcription Factors

Fred Dick and Nicholas Dyson*

Abstract

E2F activity is largely controlled by cell cycle dependent phosphorylation of the retinoblastoma family of proteins (e.g., pRB). Regulation of E2F transcription factors by RB-family proteins is crucial to the regulation of cell cycle entry. In addition to masking E2F activation of transcription, pRB family proteins have been implicated in nucleating transcriptional repressor complexes containing E2F transcription factors on cell cycle regulated promoters. More recently E2F transcription factors have been shown to regulate the activity of cellular processes in differentiation and apoptosis in a manner that is independent of cell cycle control. These recent findings have revealed that E2F transcription factors participate in noncell cycle regulatory mechanisms.

Introduction

The E2F transcription factor was first identified for its role in activating transcription of adenoviral genes expressed from the E2 promoter.[1,2] Expression of E2 gene products is essential for viral DNA replication. Subsequently, it was discovered that E2F activity is controlled in a cell cycle dependent manner by the retinoblastoma protein and this has firmly established its role as a key cell cycle regulator.[3]

In addition to the many reports that have investigated E2F activity in cell cycle regulated transcription, its many varied roles in other aspects of cell physiology are also coming to light. Molecular analyses have revealed that E2F is a family of heterodimeric transcription factors.[4,5] Aside from the recently discovered E2F7 protein, E2Fs are thought to be composed of one of six E2F family proteins bound to one of two DP proteins (see Fig. 1). Generation of gene-targeted null mouse strains for six E2F family proteins and one of the two DP proteins has shown a wide variety of phenotypes indicating that the essential functions of the individual components are quite different from one another.[6] Analysis of E2F family proteins in chromatin immunoprecipitation experiments has also broadened the quantity and types of transcriptional targets that are regulated by E2Fs.[7-9]

This review of E2F function will examine the current models of E2F activity in cell cycle control, but will also highlight many of the exciting new insights into E2F activity in development and apoptosis. These two aspects of E2F function are particularly noteworthy because they represent functions that are clearly distinct from cell cycle control and are not merely the regulation of the cell cycle in specialized cell types or under unique circumstances.

*Corresponding Author: Nicholas Dyson—Laboratory of Molecular Oncology, MGH Cancer Center, 149 13th Street, Charlestown, Massachusetts 02129, U.S.A. Email: dyson@helix.mgh.harvard.edu

Rb and Tumorigenesis, edited by Maurizio Fanciulli. ©2006 Eurekah.com and Springer Business+Science Media.

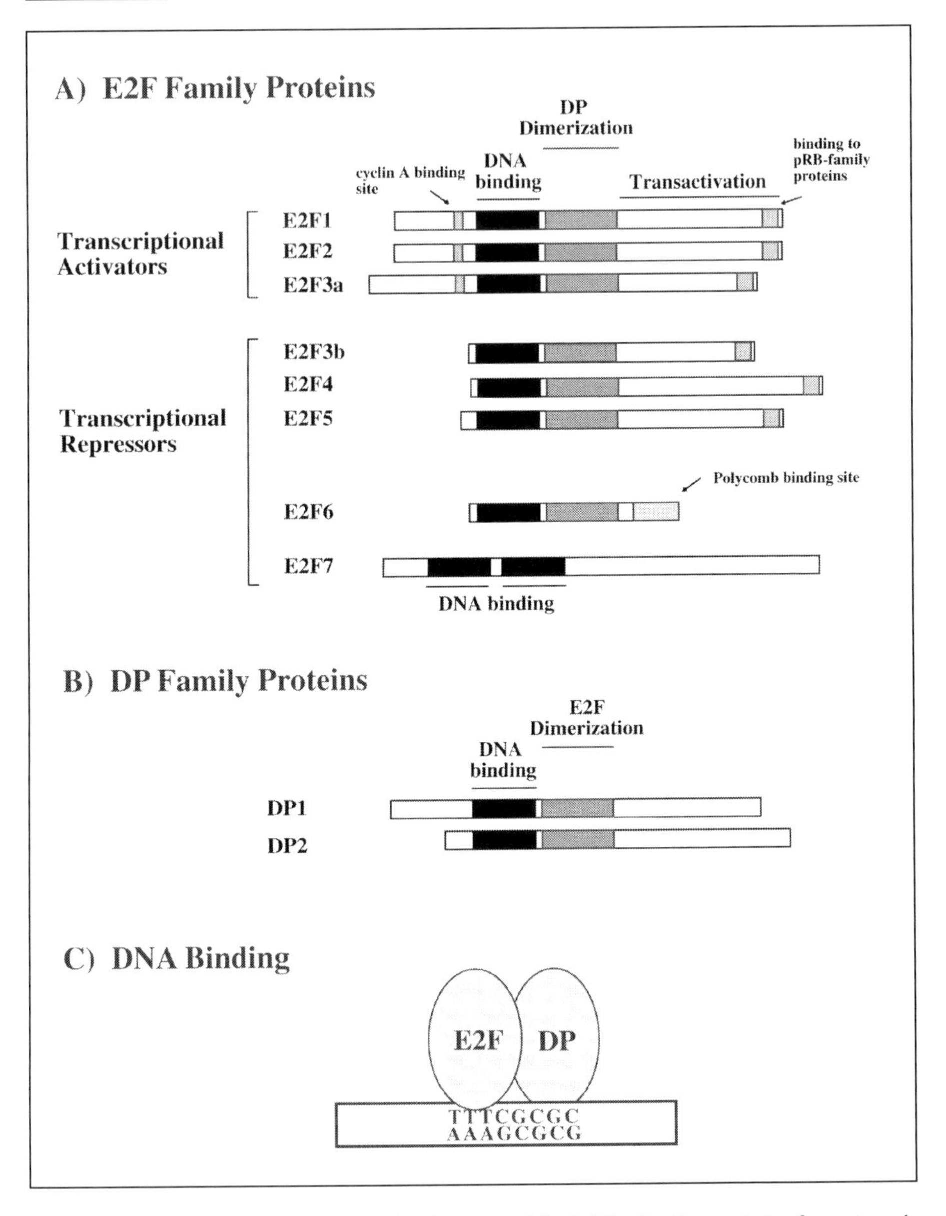

Figure 1. Structure and composition of E2F family proteins. The E2F family of transcription factors is made up of seven members, three activators and four repressors. Through the use of alternative transcriptional start sites there are 8 distinct polypeptides. Each contains a DNA binding domain, a DP dimerization domain, and E2Fs 1-5 also contain a C-terminal RB-family protein binding domain. In addition, E2F1, 2, and 3a all have a cyclin A binding site in their N-terminus. E2F6 has a binding domain to interact with polycomb proteins instead of RB-family proteins (see panel A). All E2F transcription factors are dimeric molecules that contain an E2F and a DP subunit except for E2F7 which possesses two intrinsic DNA binding domains and does not bind DP proteins. There are two DP family proteins that both contain DNA binding and E2F dimerization domains (see panel B). Dimeric E2F/DP proteins stabily associate with double stranded DNA bearing the TTTCGCGC consensus or similar sequences (see panel C).

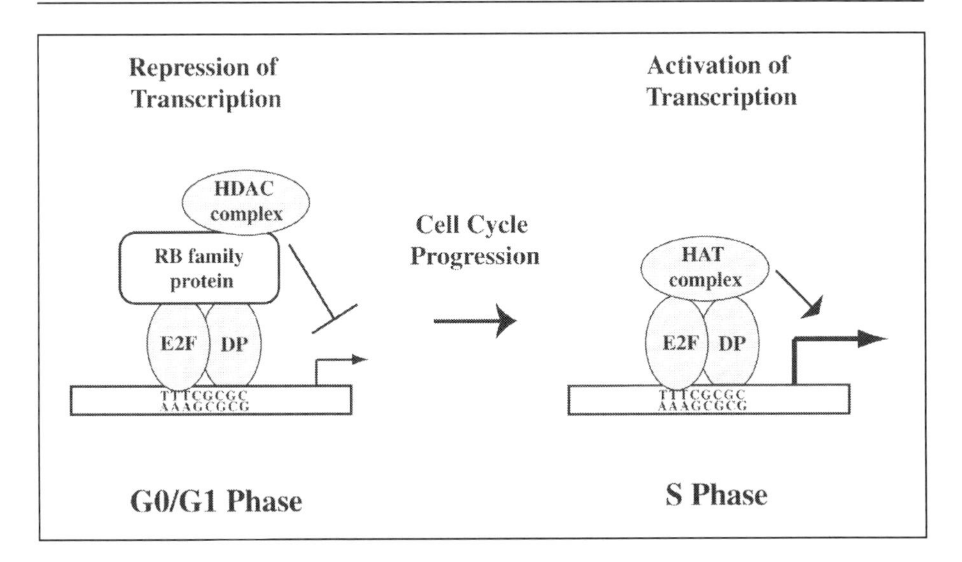

Figure 2. E2F mediated regulation of gene transcription. In quiescent and early G1 cells transcription of E2F site-containing promoters is turned off by repressor complexes nucleated by E2F transcription factors, an RB family protein, and a chromatin remodeling complex to condense the promoter structure. These repressor complexes usually contain one of E2F3b, 4, or 5. In late G1 or early S-phase the repressor complex is removed and replaced with an activator E2F like E2F1, 2, or 3a. These E2F transcription factors in turn recruit positive regulators like histone acetyltransferases to relax the chromatin structure and activate transcription.

E2F Function in Cell Cycle Regulation

Transcriptional Control of Cell Cycle Progression

E2F transcription factors have been proposed to play crucial roles in both the stimulation of cell cycle entry and in the maintenance of quiescence. The principal mechanism underlying this dual function is the participation of different family members. E2F1, -2, and -3 are all potent activators of transcription and through genetic analysis their collective function has been argued to be essential for stimulating entry into S-phase.[4,5,10] E2Fs induce cell cycle entry by activating transcription of genes that are necessary for cell cycle progression (see Fig. 2). These include enzymes necessary for DNA synthesis like dihydrofolate reductase, thymidylate synthase, ribonucleotide reductase, and DNA polymerase alpha. E2Fs drive the expression of cell cycle regulators such as cyclin E and cyclin A, and also activate their own expression. Self induction of E2F activity is thought to drive cells irreversibly forward into S-phase because it amplifies the E2F signal to cell cycle regula-tors like cyclin E that augment progression through S-phase. In addition to genetic loss-of-function studies that demonstrate the necessity of activator E2Fs for cell cycle progression,[11] ectopic expression of activator E2Fs is sufficient to induce the expression of many of the above mentioned genes and to induce DNA synthesis in quiescent cells.[10,12] These experiments have demonstrated that E2F transcription factors are potent inducers of cell cycle entry.

In contrast to the activator E2Fs, E2F4, -5, and -6 have all been proposed to function in a negative manner to restrict entry into the S-phase of the cell cycle.[4,5] Their mechanism of action is proposed to occur through the nucleation of transcriptional repressor complexes on the promoters of cell cycle regulated genes. Using gene-targeted knock-out mice it has been demonstrated that E2Fs 4 and 5 are necessary for induction of a cell cycle arrest following ectopic expression of p16 or dominant negative Ras.[13] Using chromatin immunoprecipitations it has also been shown that during quiescence the promoters of E2F responsive genes are occupied by E2F4 and -5 transcription factors, pRB family proteins and chromatin remodeling enzymes.[14-16] Upon stimulation to enter the cell cycle, these repressive complexes are replaced by activator E2Fs.[16] Chromatin immunoprecipitations with antibodies directed at acetylated histone tails, and histone deacetylase enzymes have revealed that the promoters of E2F target genes are largely deacetylated in G0 and early G1 and are occupied by histone deacetylases, in late G1[14] histones become acetylated and gene transcription is stimulated.[16,17] These experiments indicate that E2F target genes are repressed through chromatin remodeling during quiescence that is caused by histone deacetylation (see Fig. 2). Likewise, activator E2Fs have been demonstrated to bind to histone acetyltransferases like p300 to acetylate histones and induce gene expression.[18-20] E2F6 uses a different mechanism of chromatin remodeling to maintain quiescence. E2F6 lacks the conserved RB family binding domain at its C-terminus. Instead, it interacts with a complex of polycomb group proteins to establish a repressive chromatin structure on E2F responsive promoters.[21,22] This complex has been proposed to occupy these promoters during quiescence. Recently a seventh E2F family protein has been described and it has been proposed to negatively regulate transcription.[23-25] E2F7 has been suggested to repress a subset of E2F-regulated promoters. Unlike other E2Fs, E2F7 contains two DNA-binding domains, but lacks the ability to interact with DP proteins. The mechanism of E2F7-mediated repression is unknown.

The above mentioned experiments have demonstrated that members of the E2F family of transcription factors have opposing roles in cell cycle regulation. In order to better understand the interaction between these opposing activities, investigators have exploited the smaller E2F gene family in the fruit fly *Drosophila melanogaster*. Fruit flies have only two E2F transcription factors, one activator called dE2F1 and a repressor called dE2F2.[26] Similar to mammalian experiments, cell cycle progression is inhibited by the loss of dE2F1 activity.[27] When combined with a loss of dE2F2 function, the cell cycle arrest found in early embryos is rescued and development occurs up to the mid or late pupal stage.[26] Interestingly, cell cycle progression in the absence of all dE2F proteins is much slower.[28] Taken together, these experiments indicate that E2F control of the cell cycle is a product of competing positive and negative activities that may not be absolutely essential for the completion of S-phase but critical to the normal cell cycle progression that occurs in development. Based on these experiments it is formally possible that defective cell cycle entry in E2F1, 2, and 3 knock-out cells discussed above is the product of a E2F4 or 5 gain of function.[11]

Cell Cycle Control of E2F Activity

While E2F activity is important in regulating the advancement of the cell cycle, it is noteworthy that E2F activity is itself also regulated by the cell cycle. This regulation occurs on a number of levels. Serum stimulation of cell growth induces cyclin dependent kinase activity that targets the RB family of proteins (pRB, p107, and p130).[4] In particular this leads to the disruption of p130/E2F4 and p130/E2F5 complexes and pRB/ E2F3b and pRB/ E2F4 complexes from E2F responsive promoters. Cyclin/cdk phosphorylation of pRB family proteins targets the flexible spacer and C-terminal regions.[29,30] Phosphate groups in these regions are proposed to induce conformational changes that preclude E2F binding to the pocket domain.[31,32] Thus, cell cycle entry signals serve to disassemble E2F complexes that provide negative cell cycle regulation. Cell cycle dependent phosphorylation also releases E2F1, 2, and 3a from pRB to activate transcription of S-phase genes. Based on this experimental data it has been proposed that the majority of E2F regulation occurs through its physical interaction with pRB family proteins.[4]

Outside of regulation by RB family proteins, activator E2Fs also serve to auto regulate themselves by stimulating their own synthesis in S-phase.[4,5] Increasing the abundance of activator E2F proteins during a phase of the cell cycle when pRB family proteins are inactive due to phosphorylation results in elevated levels of transcription. At the end of S-phase the abundant E2F transcriptional activity is down regulated by cyclin A/cdk 2 phosphorylation.[33,34]

This regulatory event serves to block DNA binding by these phosphorylated transcription factors, in addition protein degradation by proteasomes targets E2F1 for destruction and thus reduces its transcriptional activity specifically at the S/G2 boundary.[35]

Summary

Research into the E2F family of transcription factors has revealed a wealth of information on how cell cycle progression is regulated. From this work we now have a very detailed picture of how pRB family proteins interact with E2F transcription factors and the functional consequences of these interactions. In addition, chromatin immunoprecipitation experiments have recently provided great insight into the sequence of events that occur at E2F responsive promoters during the crucial decision making steps leading to cell cycle entry.

E2F in Development

Control of Developmental Processes by E2F Family Proteins

Numerous examples of arrested development have been described in organisms that are defective for E2F function. Most notably, *Drosophila* embryos deficient for dE2F1 have very low levels of DNA synthesis and arrest in development before the end of the larval stage.[27] In mouse development, *Dp1* deficiency has been shown to prevent endocycles in giant trophoblast cells.[36] The failure of DNA synthesis in these cells results in a placental defect that ultimately leads to the death of the embryo. Because E2F is known to promote the expression of genes that are needed for cell cycle progression, it was expected that E2F proteins would be required for cell proliferation during animal development. However, many E2F mutant animals have tissue specific defects that do not obviously result from changes in cell cycle control or abnormal proliferation, and these changes reveal that E2F proteins have additional functions that are distinct from cell cycle control. For example, *E2f3*$^{-/-}$ mice have heart defects,[37] *E2f1*$^{-/-}$ animals show testicular atrophy,[38] *E2f6*$^{-/-}$ mice have defects of the axial skeleton.[39] A more extensive description of the phenotypes of E2F knockout mice has been reviewed by DeGregori.[6] Unfortunately studying development in small mammals is impeded by the difficulties associated with intra-uterine development. It is often difficult to identify the molecular basis for mammalian developmental defects and establishing the relationship between most E2F phenotypes and cell cycle control in the knockout mouse remains a formidable challenge.

Recent examples of E2F function in development that appear to be independent of cell cycle have been reported in lower organisms whose early development is easier to study. These defects include loss of asymmetry in cell division,[40] homeotic transformations,[41] defects in cell specification,[42] and the misexpression of sex specific genes.[9] These examples differ from mere cell cycle regulation because they cause cell fate decision defects in cells that are continually proliferating in early embryos or that continue to proliferate after the specification event occurs. (See Fig. 3 for models of E2F function in development).

Defective Cell Fate Specification

During the early development of *C. elegans* embryos, the SKN1 protein is preferentially localized to the posterior cell in two cell embryos.[43] Mislocalization of SKN1 to the anterior cell results in inappropriate muscle specification, ectopic anterior mesoderm, and results in lethality. Mutations in either *efl-1* or *dpl-1*, the *C. elegans* orthologs of E2F and DP genes, can produce embryos that mislocalize SKN1 and display this phenotype.[40] Normal development can be restored by combining *efl-1* or *dpl-1* mutants with loss of function mutants in *let-60*, *lin-45*, and *sem-5*, that each encode components of the Ras/MAPK pathway. These results indicate that Ras signaling opposes the activity of EFL1 early in *C. elegans* development.

The EFL1 protein is highly expressed in a region of the *C. elegans* gonad that is enriched for cells at the pachytene stage of oocyte development.[40] Its expression is undetectable in oocytes or early embryos where the asymmetry phenotype is found. In the absence of the EFL1 protein, MAPK activation is increased throughout the gonad including in mature oocytes.

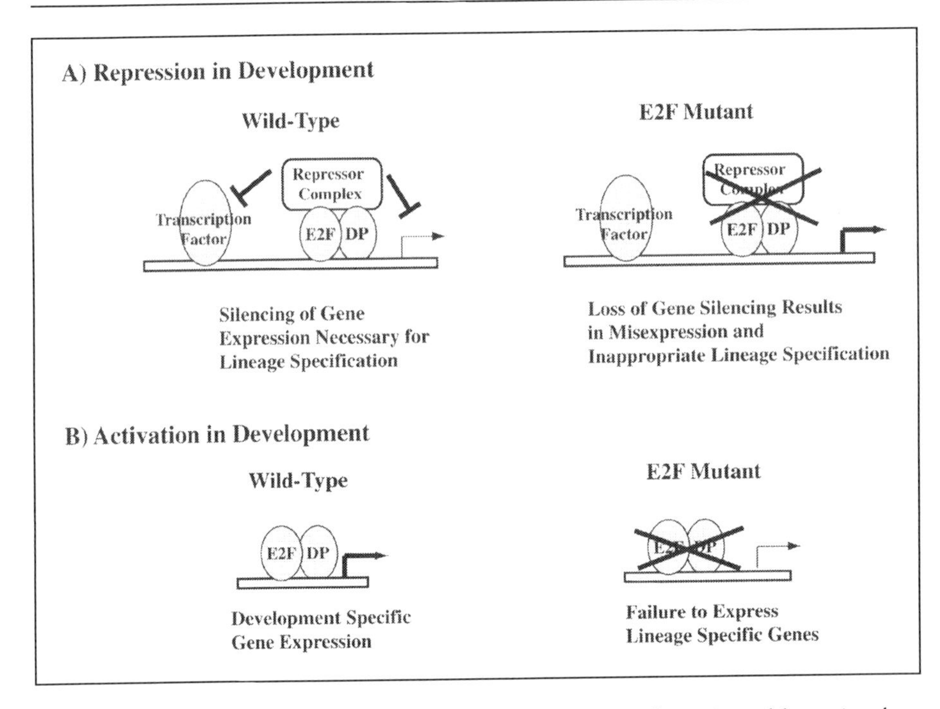

Figure 3. Mechanisms of E2F activity in development. E2F transcription factors in model organisms have been shown to play crucial roles in the specification of cell fates and patterning. E2Fs can function in development by recruiting repressors to silence transcription and prevent the expression of genes that will alter cell specification or developmental patterning. In the absence of E2F activity these repressors are missing allowing activation of transcription and misexpression to occur (see panel A). In some cases activation of transcription by E2F family proteins has been proposed to drive expression of lineage or pattern specific genes. In the absence of E2F activity patterning can be incorrectly specified (see panel B).

Normally, MAPK activity is low in developing oocytes and is not induced until fertilization. These data suggest that EFL1 regulates genes that serve to activate the Ras signaling pathway during oocyte maturation and this is the underlying cause of the later asymmetry defect.

Another developmental event in *C. elegans* that requires proper functioning of E2F transcription factors is the formation of the vulva. During the first and second larval stages (L1 and L2) six ectodermal blast cells present on the ventral midline of the developing worm are selected to acquire the vulval fate.[44] At this stage unspecified cells will fuse with the hypodermal syncytium and will no longer be able to adopt the vulval fate. Three of these cells, P(5, 6, and 7).p, are induced in the third larval stage to proliferate and become the 22 cells necessary for forming the adult vulva. The remaining three cells that were selected in L1 and L2, P(3, 4, and 8).p, now fuse with the hypodermal syncytium. Activating mutations in the Ras/MAPK pathway cause these three extra cells to remain competent for vulva formation and this results in a multivulva phenotype (Muv).[45]

Mutations or loss of function of *efl-1*, *dpl-1*, *lin-35* (a pRB family protein), *lin-53* (a component of the NuRD chromatin remodeling complex, and *hda-1* (histone deacetylase) are unable to confer the Muv phenotype on their own.[42,46] Instead, these mutations only cause a multivulva phenotype when combined with a mutation in another genetic pathway. Thus the E2F, RB, and NuRD mutants are considered class B synthetic multivulva genes (synMuv B), while the other pathway is referred to as synMuv A. Presently very little is known about the synMuv A class of genes. None of the currently identified genes in this class show homology to genes of known function. Recently a third class of genes has been described synMuv C, that shows synthetic interactions with both synMuv A and synMuv B.[47]

The synMuv B group of genes encodes proteins that include the likely orthologs of components of the dREAM and NuRD complexes.[47] dREAM and NuRD are two chromatin associated complexes that have been shown to contain, or cooperate with, RB/E2F proteins in transcriptional repression in Drosphila[48] and mammalian cells.[49] The fact that these *C. elegans* mutants share a similar pattern of genetic interactions strongly suggests that the RB/E2F function that is critical in vulva formation is one of transcriptional repression (see Fig. 3A). Intriguingly, mutations in the synMuv B group of genes result in lower expression of a *lin-39::lacZ* reporter gene in P5.p and P6.p cells.[50] *Lin-39* is a Hox gene that serves to regulate cell fusion with the hypodermal syncytium. This suggests that the class B genes may regulate activation of *lin-39* expression as well as repress others. It is also noteworthy that, similar to the asymmetry defects described above, the Muv phenotype is sensitive to changes in both the Ras/MAPK and RB/E2F pathways. In both phenotypes Ras/MAPK and RB/E2F appear to act antagonistically.

Patterning Defects in Development

A *Xenopus laevis* E2F family member (called xE2F) was identified in an expression cloning screen for genes that dominantly alter anterior-posterior patterning.[41] In comparison with human E2F family proteins the xE2F protein sequence most strongly resembles human E2F3, with the next best matches being to E2F2 and E2F1 suggesting that xE2F is a transcriptional activator. In this screen, ectopic expression of cDNAs in ectodermal explants were used to find genes that induce posterior specific homeobox genes. Indeed, the over-expression of xE2F leads to misexpression of the posterior specific HoxB9 gene in anterior structures.[41]

Construction of a chimeric gene that fuses xE2F to the *Drosophila* engrailed transcriptional repressor was used to assess loss of xE2F driven transcription. This fusion protein blocked expression of HoxB9 and disrupted normal development of the posterior region of the embryo. Injection of this chimeric RNA into dorsal or ventral sides of early embryos gave different phenotypes in the trunk region suggestive that xE2F is also important for ventral developmental fates.

Based on the homology between xE2F and other E2F family proteins from humans it seems likely that xE2F is a transcriptional activator. In addition, conditional activation of xE2F and fusion with engrailed to generate a dominant negative provide further evidence for this interpretation. Simultaneous injection of chimeric xE2F::engrailed message with excess wild-type xE2F message overcame the dominant negative effect further arguing that the chimeric dominant negative is specific (see Fig. 3B).

One clear example of E2F function in mammalian development that affects patterning comes from the *E2f6* knock-out mouse.[39] As mentioned earlier, E2F6 is divergent from other E2F family proteins because it lacks the pRB family interaction domain at its C-terminus (see Fig. 1). E2F6 has been shown to bind to RYBP, a known polycomb component, by two-hybrid and to copurify with polycomb proteins. The *E2f6* knock-out mouse does not seem to have any defects in cell cycle control in primary fibroblast cultures. Likewise, homozygous mutant animals are born at the expected mendelian frequency indicating that there are no defects that lead to early lethality. The only phenotypes reported to date from these animals are homeotic transformations of the anterior-posterior axis. In *E2f6* null animals the sixth lumbar vertebra is transformed to resemble the first sacral vertebra and the thirteenth thoracic vertebra shows degeneration of its associated ribs and now resembles the first lumbar vertebra. This phenotype is reminiscent of the transformations seen in a polycomb mutant mouse where the *Bmi1* gene is deleted.[51]

Biochemical isolation of polycomb and E2F6 complexes, coupled with the similarity in homeotic phenotypes of *E2f6* and *Bmi1* null animals it seems logical that this repressor complex plays a role in anterior-posterior patterning (see Fig. 3A). It is not clear how the lack of a cell cycle phenotype in *E2f6*$^{-/-}$ cells should be reconciled with chromatin immunoprecipitation data that indicates E2F6/polycomb proteins are present at cell cycle target genes in quiescent cells.[22] One potential explanation is that the role of E2F6 is functionally redundant with other repressor E2Fs at cell cycle targets, but E2F6 may be the only family member able to act at genes involved in anterior-posterior patterning.

Sex Specific Gene Expression

Drosophila melanogaster have two E2F family proteins, one each in the activator and repressor classes of E2Fs. Using RNAi to knock-down expression of individual *Drosophila* E2F and RB-family proteins alone or in combination, Dimova et al demonstrated by DNA microarray analysis that there are multiple classes of E2F target genes that can be grouped based on their RB/E2F regulatory characteristics.[9] Of particular note is a class of genes that are not reduced by deficiency of the activator dE2F1, that become up-regulated when the repressor dE2F2, or RB-family proteins are missing. This expression pattern suggests that E2F regulates these genes by repression.

Chromatin immunoprecipitations confirmed that RB/E2F repressors occupy the promoters of these repression specific genes.[9] Curiously none of these E2F repression-only genes are known cell cycle regulators. Northern blot analysis of RNA from male and female *Drosophila* reproductive organs has revealed that restricted expression patterns are lost in *dE2f2* mutants.[9] In some cases sex specific expression patterns are reversed between males and females. These genes are also normally repressed in proliferating *Drosophila* S2 cells. Analysis of cells synchronized in S-phase demonstrates that the RB/E2F repressor modules that regulate these genes are resistant to the inactivating activities that accompany cell cycle progression and accentuates the noncell cycle nature of this E2F mediated repression mechanism.

Summary

The role of E2F transcription factors in noncell cycle functions during development appears to be quite varied from organism to organism. This could reflect the differences in the overall developmental programs of these organisms or the differences in complexity of their RB and E2F gene families. Yet some parallels between these different developmental paradigms do emerge. One unifying theme discussed here is the frequency with which E2F and RB family mutations affect reproductive tissues. These emerge in two instances with *C. elegans* and one in *Drosophila*.[9,40,47] Also noteworthy, are the histological defects in male reproductive tissues of *E2f1* and *E2f6* knock-out mice.[38,39] Taken together these observations highlight the involvement of E2F activity in shaping the development of reproductive tissues, and we suggest that this aspect of E2F function may be highly conserved during evolution.

In all but the *Xenopus* studies, the E2F mediated events in development appear to require transcriptional repression. In *C. elegans* and *Drosophila*, this repression is most likely mediated by RB family proteins. In contrast, E2F6 mediated repression appears to be independent of RB family proteins and it will be interesting to discover how the mechanism of action of E2F6-recruited repressors compares with the repressors recruited by pRB-related proteins.

Another parallel among the examples here is the antagonism between RB/E2F function and the Ras/MAPK signaling pathway. Cell culture studies using Rb knock-out fibroblasts have previously shown Ras activity to be deregulated following loss of Rb function.[52] The original report of this phenomenon suggested that repression of gene expression is key for RB regulation of Ras and the induction of differentiation. Reduced Ras activity by mutating the N-Ras gene rescues the myogenic defect in Rb-/- embryos, suggesting that there may be antagonism with repressive RB-E2F complexes in mammalian development too.[53]

The current literature on E2F function in development raises as many questions as it answers, yet the results discussed here represent general themes that run through multiple studies of RB/E2F proteins in different model systems. These themes may provide useful paradigms for future studies. We note that the examples of developmental defects caused by mutations in E2F transcription factors described here are likely just the beginning to this newly opening field.

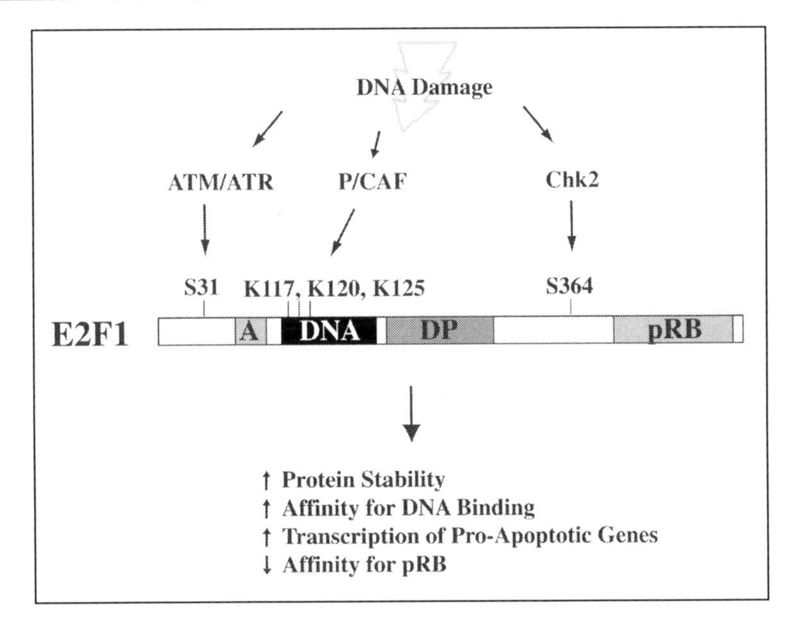

Figure 4. DNA damage induced activation of E2F1. In proliferating cells irradiation or DNA damaging drugs act to generate signals that activate E2F1. These signals include ATM/ATR phosphorylation of serine 31, Chk2 phosphorylation of serine 364, and pCAF acetylation of lysines 117, 120, and 125. These modifications increase protein stability, increase E2F1 affinity for its consensus DNA binding element, and increase transcription of pro-apoptotic genes. These modifications likely antagonize E2F1 binding to pRB under these conditions.

The Stress Response by E2F

Regulation of E2F Activation

The ability of DNA damaging agents to induce a cell cycle arrest has been known for some time. Not surprisingly DNA damage induced cell cycle arrest is accompanied by dephosphorylation of pRB and the repression of E2F regulated transcription.[54] Interestingly, the level and activity of E2F1 has recently been shown to be induced by DNA damaging agents implying that it is activated simultaneously with pRB.[55,56] Thus DNA damage appears to provide dual E2F regulating activities that inhibit proliferation through E2Fs while activating E2F1 for another purpose. Activation of E2F1 under nonproliferative conditions indicates that its DNA damage induced function is noncell cycle related.

Evidence that E2F1 is needed for functions that are fundamentally different from the other E2Fs was provided by studies of the *E2f1*$^{-/-}$ mice.[38,57] These animals appear to develop normally, but have hypercellular spleens, thymuses, and lymph nodes presumably due to defects in apoptotic induction. Importantly, thymocytes from *E2f1*$^{-/-}$ mice are resistant to DNA damage induced apoptosis.[58] These mice later succumb to a range of tumor types including reproductive tract sarcomas and lymphomas. This data underscores E2F1's dispensibility for regulating cell proliferation in development and demonstrates it's important role in the induction of apoptosis. Additionally, *E2f1*$^{-/-}$ embryonic fibroblasts are resistant to Myc induced apoptosis while *E2f2*$^{-/-}$ and *E2f3*$^{-/-}$ cells are instead resistant to Myc induced proliferation.[59] Taken together these data suggest that in response to DNA damage E2F1 may induce apoptosis and prevent cancer. This interpretation is consistent with *E2f1*$^{-/-}$ mice being tumor prone.

Current literature indicates that DNA damage signals converge on E2F1 through at least three signal transduction pathways (see Fig. 4). The first to be described was an ATM or ATR dependent phosphorylation on serine 31 near the N-terminus of E2F1.[58] Similarly the Chk2 kinase has been reported to phosphorylate serine 364,[60] and pCAF acetylates lysine residues 117, 120, and 125.[61] Each of these post translational modifications has been shown to be associated with an increased half-life of the E2F1 protein. Acetylation also increases E2F1's affinity for its cognate DNA binding site and may explain why it is specifically targeted to the P1 p73 promoter.[61-63] In addition, E2F1 transcription factors have a reduced affinity for recombinant pRB following induction by DNA damage.[64] This observation implies that modifications on E2F1 specifically antagonize the interaction with pRB since this effect is not seen with E2F4 binding to pRB.

Stress activated protein kinases (SAPK) have also been demonstrated to function in regulating E2F activity. JNK1 has been shown to phosphorylate E2F1 and this post-translational modification inhibits DNA binding.[65] A physiological circumstance for this regulation has not been identified, however JNK1 is known to be activated by UV induced DNA damage. Intriguingly, Chk2 phosphorylation of E2F1 appears to be specific to DNA damage types and is not stimulated by UV.[60] Perhaps different types of DNA damage activate separate pathways that converge on E2F1 to differentially regulate its transcriptional activity. Another example of SAPK regulation of E2F1 is through p38 activation of E2F transcription. This involves phosphorylation of pRB and subsequent release of E2F transcription factors.[65] Phosphorylation by p38 occurs on distinct sites of pRB from the cyclin/cdk sites indicating that this regulatory pathway is separate from cell cycle control.[66] E2F induction by p38 has been demonstrated to occur in response to Fas receptor stimulation on Jurkat T cells. Presumably Fas signaling in T cells utilizes p38 to activate E2F1 induced cell death although E2F1 has not been formally shown to be required for this cell death pathway.

In post-mitotic neurons RB-E2F regulation has also been shown to participate in regulating cell death.[67] DNA damage and other stresses have been shown to activate cyclin/cdk complexes.[68] This activation appears not to be cell cycle related since these neurons have differentiated and are no longer capable of mitosis. Stimulation of this pathway leads to phosphorylation of pRB and release of E2Fs that induce apoptosis. Based on the above mentioned pathways, it appears that multiple signaling pathways that are stimulated by different exogenous insults can all converge to regulate E2F activity. Based on current literature these regulatory pathways appear to be very cell type and stimulus specific.

Mechanisms of Apoptotic Induction

The ability of ectopic E2F1 over-expression to induce apoptosis has been observed by many investigators.[69] The specificity of this effect is largely reserved for E2F1 as other E2F family proteins have much lower activity in apoptotic induction, even when expressed at comparable levels.[10] These observations have stimulated considerable research into the mechanistic regulation of apoptosis by E2F1. Many different paradigms describing E2F1 induced apoptosis have emerged and are enumerated below.

Many reports of E2F1 induced apoptosis have focused on its ability to induce transcription of pro-apoptotic genes (Fig. 5). Among these targets are the p53 related protein p73, the p53 stabilizing molecule p14 Arf (p19 in mice), Apafl a caspase activating protein, and the genes encoding Caspases 3, 7, 8, and 9.[70-73] It should be noted that elevated gene transcription of at least the caspase genes is thought to prime cells for an apoptotic death, but is not directly death inducing. This group of targets suggests that E2F1 can utilize the mitochondrial apoptotic pathway and ultimately activates caspases through both p53 dependent and independent mechanisms.

E2F1 induced cell death has been measured in vivo by analyzing *Rb1*$^{-/-}$ embryos. These embryos display marked apoptosis in the central and peripheral nervous systems and this phenotype can be rescued by crossing in null alleles of *E2f1*.[74] Using mouse strains that are deficient for p53, Apafl, and p19 Arf it has been shown that these genes are required for some forms of E2F1 dependent cell death.[75-77] This indicates that these molecules are not used by E2F1 in a single linear pathway, but rather are required for at least some separable mechanisms.

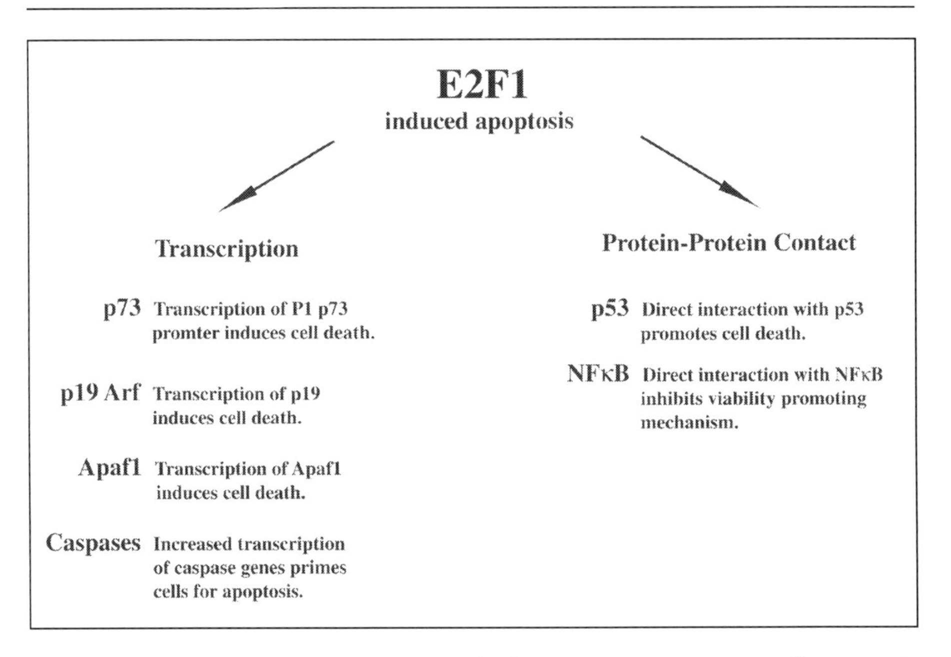

Figure 5. Mechanisms that E2F1 uses to sensitize and induce apoptosis. E2F1 sensitizes cells to apoptotic stimuli and induces cell death by a number of mechanisms. Most notably E2F1 uses its ability to activate transcription to activate expression of pro-apoptotic genes. These include p73, p14 Arf, Apaf1, and Caspases. In addition, E2F1 can directly interact with other proteins to induce cell death like p53 and the p65 subunit of NFκB.

These experiments suggest that E2F1 induces cell death by a number of different mechanisms and that these death pathways vary among tissue or cell types.

Besides the transcriptional induction mechanisms, E2F1 has also been proposed to effect cell death by direct protein-protein interaction mechanisms (Fig. 5). It has been shown that E2F1 can physically interact with p53 through its cyclin binding domain and that this interaction is sufficient for induction of cell death.[78] More recently E2F1 cell death in fibroblasts has been shown to occur by E2F1 signaling to p53 in a phosphorylation dependent manner.[79,80] This signaling event was shown to be independent of p19 Arf, suggesting that it is not a transcriptional effect. Activation of p53 under these circumstances was shown to be sensitive to caffeine suggesting that it may be mediated by the ATM kinase. Precisely how E2F1 induces ATM activity is not known, however this result agrees with the genetic analysis in mice that indicates intermediaries between E2F1 and p53, like p19, are dispensible under certain circumstances. Recently, it has also been demonstrated that E2F1 can interact with p65 and prevent assembly of the NFκB complex. In this scenario E2F1 is directly preventing the activity of a viability promoting pathway.[81]

Summary

E2F1 induced apoptosis has been an intensively studied area. From this work a myriad of different signaling pathways that can effect cell death in an E2F1 dependent manner have emerged. The use of E2F activity to kill cells in response to exogenous stimuli is a reoccurring theme from these studies. Despite the intensive study of E2F in cell death its potential mechanisms of function continue to expand. Future work will need to focus on the in vivo context in which these different paradigms occur in order for the prevalence and context of these different mechanisms to be better appreciated.

References

1. Kovesdi I, Reichel R, Nevins JR. Identification of a cellular transcription factor involved in E1A trans-activation. Cell 1986; 45(2):219-228.
2. Yee AS, Raychaudhuri P, Jakoi L et al. The adenovirus-inducible factor E2F stimulates transcription after specific DNA binding. Mol Cell Biol 1989; 9:578-585.
3. Chellappan S, Hiebert S, Mudryj M et al. The E2F transcription factor is a cellular target for the RB protein. Cell 1991; 65(6):1053-1061.
4. Dyson N. The regulation of E2F by pRB-family proteins. Genes Dev 1998; 12(15):2245-2262.
5. Trimarchi JM, Lees JA. Sibling rivalry in the E2F family. Nat Rev Mol Cell Biol 2002; 3(1):11-20.
6. DeGregori J. The genetics of the E2F family of transcription factors: Shared functions and unique roles. Biochim Biophys Acta 2002; 1602:131-150.
7. Ren B, Cam H, Takahashi Y et al. E2F integrates cell cycle progression with DNA repair, replication, and G(2)/M checkpoints. Genes and Development 2002; 16(2):245-256.
8. Wells J, Yan PS, Cechvala M et al. Identification of novel pRb binding sites using CpG microarrays suggests that E2F recruits pRb to specific genomic sites during S phase. Oncogene 2003; 22:1445-1460.
9. Dimova DK, Stevaux O, Frolov MV et al. Cell cycle-dependent and cell cycle-independent control of transcription by the Drosophila E2F/RB pathway. Genes Dev 2003; 17:2308-2320.
10. DeGregori J, Leone G, Miron A et al. Distinct roles for E2F proteins in cell growth control and apoptosis. Proc Natl Acad Sci 1997; 94:7245-7250.
11. Wu L, Timmers C, Maiti B et al. The E2F1-3 transcription factors are essential for cellular proliferation. Nature 2001; 414(6862):457-462.
12. Leone GJ, DeGregori J, Yan Z et al. E2F-3 activity is regulated during the cell cycle and is required for the induction of S-phase. Genes and Development 1998; 12.
13. Gaubatz S, Lindeman GJ, Jakoi L et al. E2F4 and E2F5 play an essential role in pocket protein-mediated G1 control. Mol Cell 2000; 6(3):729-735.
14. Rayman JB, Takahashi Y, Indjeian VB et al. E2F mediates cell cycle-dependent transcriptional repression in vivo by recruitment of an HDAC1/mSin3B corepressor complex. Genes Dev 2002; 16(8):933-947.
15. Wells J, Boyd K, Fry C et al. Target gene specificity of E2F and pocket protein family members in living cells. Mol Cell Biol 2000; 20(16):5797-5807.
16. Takahashi Y, Rayman J, Dynlacht B. Analysis of promoter binding by the E2F and pRB families in vivo: Distinct E2F proteins mediate activation and repression. Genes and Development 2000; 14(7):804-816.
17. Morrison AJ, Sardet C, Herrera RE. Retinoblastoma protein transcriptional repression through Histone deacetylation of a single nucleosome. Mol Cell Biol 2002; 22:856-865.
18. Trouche D, Kouzarides T. E2F1 and E1A 12S have a homologous activation domain regulated by RB and CBP. Proc Natl Acad Sci 1996; 93:1439-1442.
19. Morris L, Allen KE, La Thangue NB. Regulation of E2F transcription by cyclin E-Cdk2 kinase mediated through p300/CBP coactivators. Nat Cell Biol 2000; 2(4):232-239.
20. Taubert S, Gorrini C, Frank SR et al. E2F-dependent histone acetylation and recruitment of the Tip60 acetyltransferase complex to chromatin in late G1. Mol Cell Biol 2004; 24:4546-4556.
21. Trimarchi JM, Fairchild B, Wen J et al. The E2F6 transcription factor is a component of the mammalian Bmi1-containing polycomb complex. Proc Natl Acad Sci USA 2001; 98(4):1519-1524.
22. Ogawa H, Ishiguro K, Gaubatz S et al. A complex with chromatin modifiers that occupies E2F- and Myc-responsive genes in G0 cells. Science 2002; 296(5570):1132-1136.
23. de Bruin A, Maiti B, Jakoi L et al. Identification and characterization of E2F7, a novel mammalian E2F family member capable of blocking cellular proliferation. J Biol Chem 2003; 278:42041-42049.
24. Di Stefano L, Jensen MR, Helin K. E2F7, a novel E2F featuring DP-independent repression of a subset of E2F-regulated genes. EMBO J 2003; 22(6289-98).
25. Logan N, Delavaine L, Graham A et al. E2F-7: A distinctive E2F family member with an unusual organization of DNA-binding domains. Oncogene 2004; 23:5138-5150.
26. Frolov MV, Huen DS, Stevaux O et al. Functional antagonism between E2F family members. Genes & Development 2001; 15(16):2146-2160.
27. Duronio RJ, O'Farrell PH, Xie J-E et al. The transcription factor E2F is required for S phase during Drosophila embryogenesis. Genes Dev 1995; 9:1445-1455.
28. Frolov MV, Stevaux O, Moon NS et al. G1 cyclin-dependent kinases are insufficient to reverse dE2F2-mediated repression. Genes Dev 2003; 17:723-728.
29. Lees JA, Buchkovich KJ, Marshak DR et al. The retinoblastoma protein is phosphorylated on multiple sites by human cdc2. EMBO J 1991; 10(13):4279-4290.

30. Hamel PA, Cohen BL, Sorce LM et al. Hyperphosphorylation of the retinoblastoma gene product is determined by domains outside the simian virus 40 large-T-antigen-binding regions. Mol Cell Biol 1990; 10(12):6586-6595.
31. Knudsen ES, Wang JYJ. Dual mechanisms for the inhibition of E2F binding to RB by cyclin-dependent kinase-mediated RB phosphorylation. Mol Cell Biol 1997; 17:5771-5783.
32. Harbour J, Luo R, Dei Santi A et al. Cdk phosphorylation triggers sequential intramolecular interactions that progressively block Rb functions as cells move through G1. Cell 1999; 98:859-869.
33. Dynlacht BD, Flores O, Lees JA et al. Differential regulation of E2F trans-activation by cyclin-cdk2 complexes. Genes & Dev 1994; 8:1772-1786.
34. Krek W, Ewen ME, Shirodkar SZ et al. Negative regulation of the growth-promoting transcription factor E2F-1 by a stably bound cyclin A-dependent protein kinase. Cell 1994; 78:161-172.
35. Marti A, Wirbelauer C, Scheffner M et al. Interaction between ubiquitin-protein ligase SCFSKP2 and E2F-1 underlies the regulation of E2F-1 degradation. Nat Cell Biol 1999; 1:5-7.
36. Kohn MJ, Bronson RT, Harlow E et al. Dp1 is required for extra-embryonic development. Development 2003; 130:1295-1305.
37. Cloud JE, Rogers C, Reza TL et al. Mutant mouse models reveal the relative roles of E2F1 and E2F3 in vivo. Mol Cell Biol 2002; 22:2663-2672.
38. Yamasaki L, Jacks T, Bronson R et al. Tumor induction and tissue atrophy in mice lacking E2F-1. Cell 1996; 85:537-548.
39. Storre J, Elsasser HP, Fuchs M et al. Homeotic transformations of the axial skeleton that accompany a targeted deletion of E2f6. EMBO Rep 2002; 3:695-700.
40. Page BD, Guedes S, Waring D et al. The C. elegans E2F- and DP-related proteins are required for embryonic asymmetry and negatively regulate Ras/MAPK signaling. Mol Cell 2001; 7:451-460.
41. Suzuki A, Hemmati-Brivanlou A. Xenopus embryonic E2F is required for the formation of ventral and posterior cell fates during early embryogenesis. Mol Cell 2000; 5(2):217-229.
42. Lu X, Horvitz HR. lin-35 and lin-53, two genes that antagonize a C. elegans Ras pathway, encode proteins similar to Rb and its binding protein RbAp48. Cell 1998; 95:981-991.
43. Bowerman B, Draper BW, Mello CC et al. The maternal gene skn-1 encodes a protein that is distributed unequally in early C. elegans embryos. Cell 1993; 74:443-452.
44. Ferguson EL, Sternberg P, Horvitz HR. A genetic pathway for the specification of the vulval cell lineages of Caenorhabditis elegans. Nature 1987; 326:259-267.
45. Sternberg PW, Han M. Genetics of RAS signalling in C. elegans. Trends Genet 1998; 14:466-472.
46. Ceol CJ, Horvitz HR. dpl-1 DP and efl-1 E2F act with lin-35 Rb to antagonize Ras signaling in C. elegans vulval development. Mol Cell 2001; 7(3):461-473.
47. Ceol CJ, Horvitz HR. A new class of C. elegans synMuv genes implicates a Tip60/NuA4-like HAT complex as a negative regulator of Ras signaling. Dev Cell 2004; 6:563-576.
48. Taylor-Harding B, Binne UK, Korenjak M et al. p55, the Drosophila ortholog of RbAp46/RbAp48, is required for the repression of dE2F2/RBF-regulated genes. Mol Cell Biol 2004; 24:9124-9136.
49. Korenjak M, Taylor-Harding B, Binné UK et al. Native E2F/RBF complexes contain Myb-interacting proteins and repress transcription of developmentally controlled E2F target genes. Cell 2004; 119:in press.
50. Chen Z, Han M. C. elegans Rb, NuRD, and Ras regulate lin-39-mediated cell fusion during vulval fate specification. Curr Biol 2001; 11(23):1874-1879.
51. van der Lugt NM, Domen J, Linders K et al. Posterior transformation, neurological abnormalities, and severe hematopoietic defects in mice with a targeted deletion of the bmi-1 proto-oncogene. Genes Dev 1994; 8:757-769.
52. Lee KY, Ladha MH, McMahon C et al. The retinoblastoma protein is linked to the activation of ras. Mol Cell Biol 1999; 19(11):7724-7732.
53. Takahashi C, Bronson RT, Socolovsky M et al. Rb and N-ras function together to control differentiation in the mouse. Mol Cell Biol 2003; 23:5256-5268.
54. Brugarolas J, Moberg K, Boyd S et al. Inhibition of cyclin-dependent kinase 2 by p21 is necessary for retinoblastoma protein-mediated G1 arrest after gamma-irradiation. Proc Natl Acad Sci USA 1999; 96:1002-1007.
55. Blattner C, Sparks A, Lane D. Transcription Factor E2F-1 is upregulated in response to DNA damage in a manner analogous to that of p53. Mol Cell Biol 1999; 19:3704-3713.
56. Hofferer M, Wirbelauer C, Humar B et al. Increased levels of E2F-1-dependent DNA binding activity after UV- or gamma-irradiation. Nucleic Acids Res 1999; 27:491-495.
57. Field SJ, Tsai F-Y, Kuo F et al. E2F-1 functions in mice to promote apoptosis and suppress proliferation. Cell 1996; 85:549-561.
58. Lin WC, Lin FT, Nevins JR. Selective induction of E2F1 in response to DNA damage, mediated by ATM-dependent phosphorylation. Genes Dev 2001; 15(14):1833-1844.

59. Leone G, Sears R, Huang E et al. Myc requires distinct E2F activities to induce S phase and apoptosis. Mol Cell 2001; 8:105-113.
60. Stevens C, Smith L, La Thangue NB. Chk2 activates E2F-1 in response to DNA damage. Nat Cell Biol 2003; 5:401-409.
61. Pediconi N, Ianari A, Costanzo A et al. Differential regulation of E2F1 apoptotic target genes in response to DNA damage. Nat Cell Biol 2003; 5:552-558.
62. Marzio G, Wagener C, Gutierrez MI et al. E2F family members are differentially regulated by reversible acetylation. J Biol Chem 2000; 275:10887-10892.
63. Martinez-Balbas MA, Bauer UM, Nielsen SJ et al. Regulation of E2F1 activity by acetylation. EMBO J 2000; 19:662-71.
64. Dick FA, Dyson N. pRB contains an E2F1 specific binding domain that allows E2F1 induced apoptosis to be regulated separately from other E2F activities. Mol Cell 2003; 12:639-649.
65. Wang S, Nath N, Minden A et al. Regulation of Rb and E2F by signal transduction cascades: Divergent effects of JNK1 and p38 kinases. EMBO J 1999; 18:1559-1570.
66. Nath N, Wang S, Betts V et al. Apoptotic and mitogenic stimuli inactivate Rb by differential utilization of p38 and cyclin-dependent kinases. Oncogene 2003; 22:5986-5994.
67. Park DS, Morris EJ, Bremner R et al. Involvement of retinoblastoma family members and E2F/DP complexes in the death of neurons evoked by DNA damage. J Neurosci 2000; 20(9):3104-14.
68. Park DS, Morris EJ, Padmanabhan J et al. Cyclin-dependent kinases participate in death of neurons evoked by DNA-damaging agents. J Cell Biol 1998; 143:457-467.
69. Ginsberg D. E2F1 pathways to apoptosis. FEBS Lett 2002; 529:122-125.
70. Irwin M, Marin MC, Phillips AC et al. Role for the p53 homologue p73 in E2F-1-induced apoptosis. Nature 2000; 407(6804):645-648.
71. Bates S, Phillips AC, Clark PA et al. p14ARF links the tumour suppressors RB and p53. Nature 1998; 395(6698):124-125.
72. Moroni M, Hickman E, Denchi E et al. Apaf-1 is a transcriptional target for E2F and p53. Nat Cell Biol 2001; 6:552-558.
73. Nahle Z, Polakoff J, Davuluri RV et al. Direct coupling of the cell cycle and cell death machinery by E2F. Nat Cell Biol 2002; 4:859-864.
74. Tsai KY, Hy Y, Macleod KF et al. Mutation of E2F1 suppresses apoptosis and inappropriate S-phase entry and extends survival of Rb-deficient mouse embryos. Mol Cell 1998; 2:293-304.
75. Macleod KF, Hu Y, Jacks T. Loss of RB activates both p53-dependent and independent cell death pathways in the developing mouse nervous system. EMBO J 1996; 15(22):6178-6188.
76. Guo Z, Yikang S, Yoshida H et al. Inactivation of the retinoblastoma tumor suppressor induces apoptosis protease-activating factor-1 dependent and independent apoptotic pathways during embryogenesis. Cancer Res 2001; 61:8395-8400.
77. Tsai KY, MacPherson D, Rubinson DA et al. ARF is not required for apoptosis in Rb mutant mouse embryos. Curr Biol 2002; 12(2):159-163.
78. Hsieh JK, Yap D, O'Connor DJ et al. Novel function of the cyclin a binding site of E2F in regulating p53-induced apoptosis in response to DNA damage. Mol Cell Biol 2002; 22:78-93.
79. Rogoff HA, Pickering MT, Debatis ME et al. E2F1 induces phosphorylation of p53 that is coincident with p53 accumulation and apoptosis. Mol Cell Biol 2002; 22:5308-5318.
80. Rogoff HA, Pickering MT, Frame FM et al. Apoptosis associated with deregulated E2F activity is dependent on E2F1 and Atm/Nbs1/Chk2. Mol Cell Biol 2004; 24:2968-2977.
81. Phillips AC, Ernst MK, Bates S et al. E2F-1 potentiates cell death by blocking antiapoptotic signaling pathways. Mol Cell 1999; 4(5):771-781.

CHAPTER 7

Regulation of E2F-Responsive Genes through Histone Modifications

Estelle Nicolas, Laetitia Daury and Didier Trouche*

Abstract

The retinoblastoma protein Rb, when targeted to E2F-responsive promoters through a direct interaction with E2F proteins, actively represses transcription. This property is shared by the two Rb-related proteins, p107 and p130. Active transcriptional repression by Rb is important for the proper control of cell growth. Many recent results have indicated that Rb represses transcription through proteins acting on chromatin structure. The purpose of this chapter is to review these results, and to discuss the possible mechanisms by which accurate regulation of E2F-responsive genes is achieved.

Introduction

The retinoblastoma protein Rb and its two cousins the p107 and p130 proteins, collectively called "pocket proteins", play a critical role in the control of mammalian cell proliferation.[1] They are active and hypophosphorylated in resting cells and at the beginning of the G1 phase of the cell cycle, where they restrained progression towards S phase. At the end of G1, their phosphorylation mediated by cyclin/cdks leads to their inactivation (Fig. 1).

One of the major targets of pocket proteins is the E2F transcription factor.[2] The E2F transcription factor is responsible for the transcriptional activation of many S phase specific genes at the end of G1 and the beginning of S phase[3] (Fig. 1). E2F binds directly to a so-called "E2F site" in the promoter of these genes. E2F is composed of a family of proteins, called E2F1 to 7. E2F 1 to 6 bind as a heterodimer with a dimerisation partner DP (there are two DP proteins, DP1 and DP2), whereas the recently described E2F7 protein[4] binds to a subset of E2F sites in a DP-independent manner (Fig. 2). E2F1 to 5 share a transcriptional activation domain at their C-terminus. This domain is absent from E2F6 and E2F7, and these two proteins function as transcriptional repressors.[4-6]

In resting cells and at the beginning of G1, pocket proteins bind directly to the transcriptional activation domain of E2F1-5, with some specificity (Fig. 2).[2] Once bound to these E2Fs, they are recruited to E2F-regulated promoters that they actively repress.[7] Phosphorylation of pocket proteins by cyclin/cdks disrupts the E2F/pocket proteins interaction, and promoter-bound E2F can then activate transcription (Fig. 1). Thus, classical E2F-responsive genes are actively repressed in resting cells and the beginning of G1 and activated at the end of G1 and the beginning of S phase (Fig. 1).

Because of the importance of Rb in cancer, the molecular mechanisms which are responsible for the correct regulation of E2F-responsive genes have been extensively studied. In vitro

*Corresponding Author: Didier Trouche—Laboratoire de Biologie Moléculaire Eucaryote, CNRS UMR 5099, Institut d'Exploration Fonctionnelle du Génome, Toulouse, France. Email: trouche@ibcg.biotoul.fr

Rb and Tumorigenesis, edited by Maurizio Fanciulli. ©2006 Eurekah.com and Springer Business+Science Media.

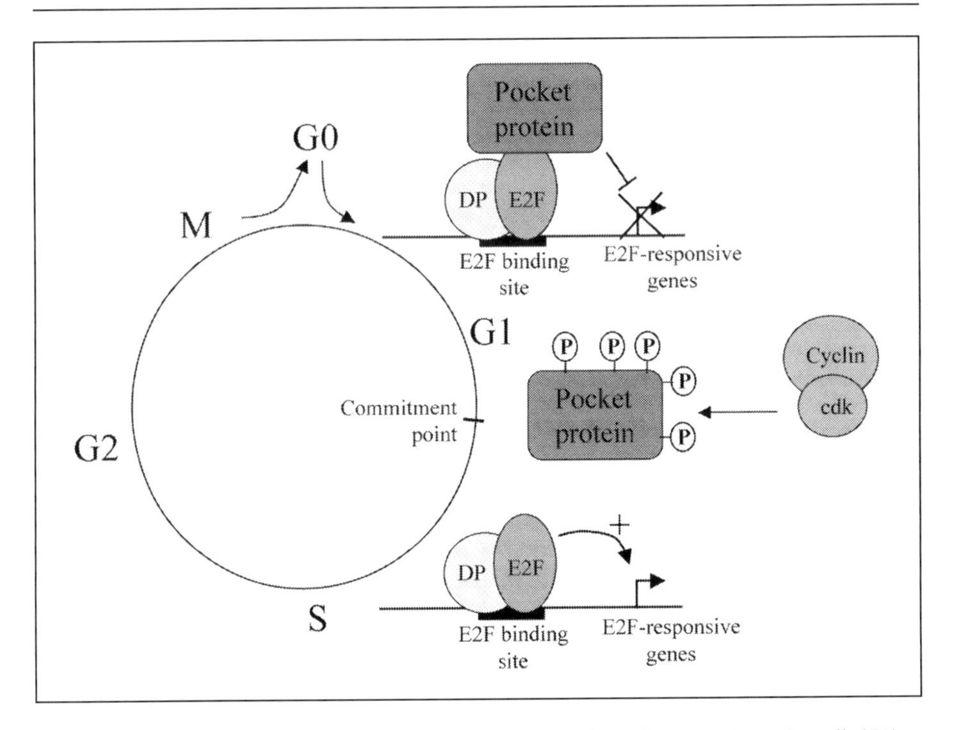

Figure 1. Cell cycle-dependent evolution of protein binding on a classical E2F site. In resting cells (G0) or cells at the beginning of G1, E2F-responsive promoters are bound by E2F transcription factor complexed with a pocket protein. This pocket protein mediates transcriptional repression. When cells progress into G1, pocket proteins are inactivated through phosphorylation by cyclin/cdks. The E2F/DP heterodimer (called "free E2F") can then activate transcription.

reconstitution of transcriptional repression by Rb has demonstrated that chromatin structure plays a critical role in these processes.[8]

In eukaryotic cells, the DNA is packaged with protein in a compacted structure called chromatin. Chromatin is the real substrate of enzymes acting on DNA, such as DNA and RNA polymerases. Therefore, chromatin structure plays an important role in regulation of transcription. The basic unit of chromatin is the nucleosome, which consists of an octamer of small basic proteins, the histones, around which 146 base pairs of DNA are wrapped. Chromatin function can be regulated through the post-translational modifications of nucleosomal histones. Most of these modifications occur within their short N-terminal tails, which protrude out of the nucleosome and are thus accessible to enzymes.[9]

Acetylation of lysines, the best characterized modification, usually correlates with transcriptional activation.[10] It has been suggested for long that through the neutralization of the positive charges of the lysines, it weakens histone/DNA or histone/histone interaction, thereby creating a more open structure, more permissive for transcription. Such a mechanism is less likely for other histone modifications, such as lysine methylation, which do not significantly change the charge of the histones. The fact that these latter modifications also affect chromatin function has led to the proposal that histone post-translational modifications could function as signals regulating interaction of specific proteins with nucleosomes. Consistent with this hypothesis, some acetylated lysines are recognised by some bromodomains,[11,12] whereas methylated lysines can function as binding sites for some chromodomains.[13,14] Both domains were defined by sequence analysis and can be found in many proteins functionally related to chromatin. The emerging notion is that the various modifications function together, in an interdependent way, to specify a given

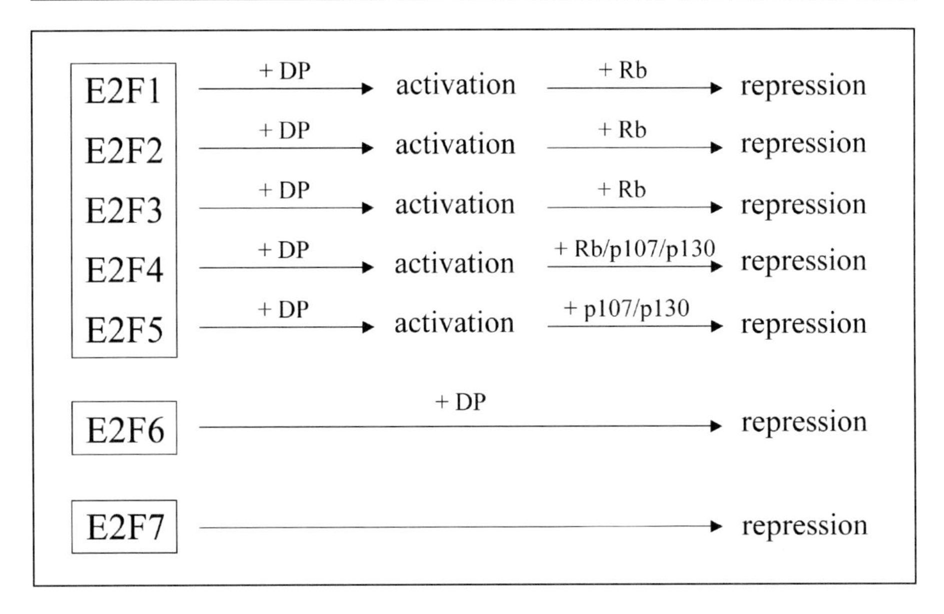

Figure 2. The E2F family in transcription. E2F1 to 5 bind and activate E2F-responsive promoters as heterodimers with a DP protein. These E2F are targeted by specific pocket proteins (either Rb, p107 or p130), which convert them into transcriptional repressors. E2F6 binds DNA as a heterodimer with DP, and functions as a repressor by itself, whereas E2F7 binds and represses a subset of E2F-reponsive promoters in a DP-independent manner.

functional state for chromatin ("histone code" hypothesis).[15] The enzymes which set up these modifications can be recruited by sequence specific transcription factors to modify locally chromatin structure, leading to the activation or the repression of transcription from their target promoters.[16] Since transcriptional repression of E2F-responsive genes is critical for Rb function, the involvement of histone modifying activities in the regulation of E2F-responsive promoters has been extensively studied in the past few years.

Regulation of E2F-Responsive Genes through the Control of Histone Acetylation

The first clue suggesting that enzymes governing histone modifications could be important for the regulation of E2F-responsive genes came when a known E2F coactivator, CBP,[17] was found to possess histone acetyl transferase (HAT) activity.[18,19] Since then, another HAT, GCN5, was also proposed to be important for activation of E2F-regulated promoters.[20] However, a major breakthrough was made when several groups found that Rb interacts with histone deacetylases (HDAC), in particular HDAC1.[21-23] A sequence within HDAC1 which can bind directly to the transcriptional repression domain of Rb was identified.[21] Subsequent studies indicated that Rb could also interact with HDAC2 and HDAC3.[24] Since HDAC3 does not contain the Rb-interacting sequence, its binding to Rb is likely to be indirect. Importantly, the availability of specific HDAC inhibitors allowed the demonstration that HDAC activity was required for Rb to repress transcription, at least on some promoters.[22] Altogether, these results have led to the proposal of a model in which transcriptional repression by Rb was mediated though the recruitment of HDAC to E2F-responsive promoters, resulting in local histone deacetylation and the subsequent repression (Fig. 3). A similar model was also drawn for transcriptional repression by the Rb-related p107 and p130 proteins.[25] In agreement with this, histones on many E2F-regulated promoters are

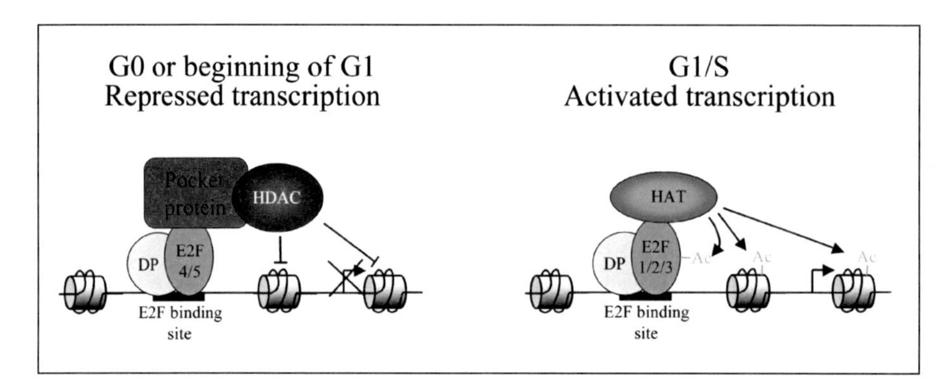

Figure 3. Regulation of E2F-responsive promoters through the sequential recruitment of HDACs and HATs. In resting cells (G0) or cells at the beginning of G1, the pocket protein bound to E2F-regulated promoters recruits a HDAC, such as HDAC1, which mediates transcriptional repression through nucleosomal histones deacetylation. Transcriptional activation at the end of G1 is mediated through the recruitment of a HAT, which acetylates both histones and E2Fs. The identity of this HAT is still unclear. Note that ChIP experiments indicate that E2F4 or 5 are bound to repressed promoters, whereas E2F1, 2 or 3 are responsible for transcriptional activation.[27]

hypoacetylated when cells are in G0 and the promoters are repressed and became hyperacetylated around the G1/S transition, when the genes are expressed[26,27] (Fig. 3). Moreover, the presence of HDACs on E2F-regulated promoters in resting cells was confirmed by chromatin immunoprecipitation experiments.[26,28] Rb does not recruit HDAC alone, but rather a histone deacetylase complex in which other subunits help the enzyme to deacetylate nucleosomes.[29] The Rb-associated complex is likely to be the previously characterized Sin3 histone deacetylase complex, since the presence of Sin3 on E2F-regulated promoters has been shown by chromatin immunoprecipitation.[28] Moreover, RbAp48, a component of the Sin3 complex, belongs to the Rb-associated HDAC complex[29] and to the CERC complex, an E2F complex involved in repression of the cyclin E promoter and which contains a pocket protein and a histone deacetylase.[30] The importance of RbAp48 and HDACs in the Rb pathway was also confirmed by genetic experiments in *C. elegans*.[31,32]

All the promoters investigated so far harboured a similar evolution from hypo- to hyper-acetylated histones during G1 progression[26,27] (Fig. 3). However, the involvement of HDACs in transcriptional repression by Rb is restricted to a subset of promoters.[22] This indicates that there are some promoter to promoter variations, and that the context of the promoter is certainly crucial for the action of histone modifying enzymes.

In addition, the enzymes which regulated histone acetylation can also modify transcription factors.[33] E2F1, 2 and 3 are acetylated by the CBP/p300 or pCAF enzymes, and their acetylation leads to increased DNA binding and transcriptional activation properties.[34,35] Strikingly, Rb itself can be acetylated by CBP/p300, and this acetylation is linked to its phosphorylation.[36] An integrated model of the involvement of HAT and HDACs in the regulation of E2F-responsive genes can thus be proposed (Fig. 3).

Methylation of Histone H3 K9

The idea that recruitment of HDACs cannot account for all the repressive activities of Rb came from in vitro studies. Whereas chromatin is important for transcriptional repression by Rb, inhibition of HDACs does not relieve Rb repression in vitro, indicating the involvement of other chromatin-based processes.[8] One of these processes is likely to involve methylation of histone H3 K9. Histone H3 K9 methylation correlates with transcriptional repression.[15] When

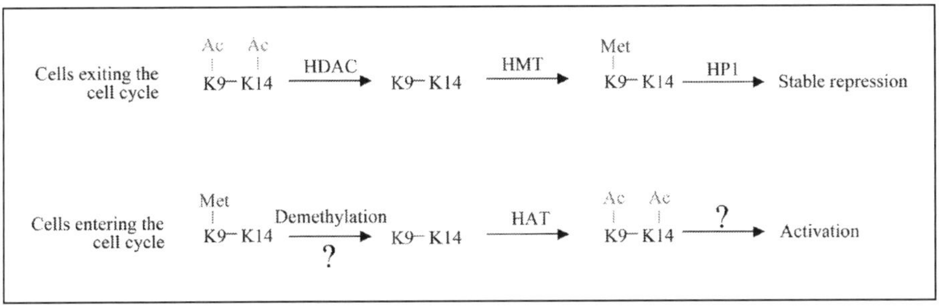

Cells exiting the cell cycle: Ac Ac K9–K14 —HDAC→ K9–K14 —HMT→ Met K9–K14 —HP1→ Stable repression

Cells entering the cell cycle: Met K9–K14 —Demethylation ?→ K9–K14 —HAT→ Ac Ac K9–K14 —?→ Activation

Figure 4. Model of the sequential evolution of histone H3 K9 and K14 modifications on E2F-regulated promoters during cell cycle exit (top) or cell cycle entry (bottom). When cells exit the cell cycle, HDACs function first to deacetylate K9 of histone H3, thereby allowing its methylation by HMTs such as Suv39H1. K9 methylation would in turn result in transcriptional repression through HP1 recruitment. When cells enter the cell cycle, K9 is demethylated.[39,41] Note that the mechanism of this demethylation is unclear, since no lysine demethylase has been described so far.[53] K9 and K14 on histone H3 are then acetylated,[39,41] as well as some lysines on histone H4,[26] leading to transcriptional activation by unknown mechanisms.

methylated, K9 is recognised by proteins of the HP1 family through their chromodomain. The importance of enzymes which methylate K9 in transcriptional repression of E2F-responsive genes is indicated by the fact that Rb, p107 and p130 physically interact with Suv39H1, a histone methyl transferase (HMT) with such a specificity.[37-39] Moreover, this interaction correlates with the ability of pocket proteins to repress transcription. Strikingly, E2F6, another repressor of E2F-responsive genes (Fig. 1), is also present within a multimolecular complex possessing HMT activity specific for histone H3 K9.[40]

Although the absence of specific inhibitors of HMTs precludes a direct demonstration of their role in E2F activity, they are likely to be important for transcriptional repression, since (i) Suv39H1 functions as a corepressor for Rb and its cousins; (ii) the transcription of some E2F-responsive genes is derepressed in cells derived from Suv39H1 knocked-out mice. Moreover, the presence of histone H3 K9 methylation on some E2F-regulated promoters correlates with their transcriptional repression by a member of the Rb protein family.[38,39,41] How does K9 methylation bring about repression? The only clue is the likely importance of HP1 proteins, which recognise the methylated K9 through their chromodomain.[13,14] First, HP1 proteins bind to the E2F-responsive cyclin E promoter when K9 is methylated and the gene is repressed.[38] Second, HP1γ, a member of the HP1 family, is present in the E2F6-associated complex.[40] Since HP1 proteins are known transcriptional repressors, which are involved in the higher order compaction of constitutive heterochromatin, their presence could induce the formation of a compacted heterochromatin-like structure, thereby resulting in stable transcriptional inactivation. Such a mechanism has recently been demonstrated to occur on E2F-regulated promoters during the process of senescence.[42]

Transcriptional repression by Rb thus requires the action of both HDACs and HMTs. Since these two enzymes modify the same substrate, they could function in a concerted fashion to induce transcriptional repression. Consistent with this possibility, they are present within the same multimolecular complex.[43,44] What could be the molecular mechanism underlying their cooperation? Strikingly, histone H3 K9 can be acetylated, and its acetylation and methylation are believed to be mutually exclusive.[45] One obvious possibility is thus that HDACs are required to deacetylate histone H3 K9, thereby allowing its methylation, HP1 recruitment , and the subsequent transcriptional repression (Fig. 4). Consistent with this possibility, H3 K9 acetylation is a prominent modification on E2F-regulated promoters when they are actively transcribed.[39,41]

Involvement of Other Proteins Functioning on Chromatin

Transcriptional repression by Rb also involves other proteins targeting chromatin. For example, hbrm, which is the catalytic sub-unit of the mammalian SWI/SNF ATP-dependent chromatin remodelling complex, functions as a corepressor for Rb.[46] The mechanism by which it cooperates with Rb to repress transcription of E2F-responsive genes is not known. However, ATP dependent chromatin remodelling machineries are known to function inter-dependently with histone modifying enzymes both biochemically and genetically (see for example ref. 47). It is thus tempting to speculate that the mammalian SWI/SNF complex participates in the establishment of the precise pattern of histone modifications corresponding to the repressed state.[48] Recently, it was shown that Rb also interacts with proteins from the polycomb group, which, in a manner reminiscent to HP1 (see above), participate in nucleating heterochromatic-like regions.[49] This result reinforces the idea that under some circumstances, transcriptional repression by Rb could involve the formation of heterochromatic-like structure on E2F-regulated promoters.

Chromatin Modifying Enzymes Involved in the E2F/Rb Pathway: Relationship with Cancer

The Rb protein, and more generally the Rb pathway, is a "hot spot" of mutations in human cancer. Because of their importance in Rb function, the enzymes and proteins described above could theoritically also be mutated in human cancers. There are some clear evidence that it is the case. The E2F coactivators CBP/p300 are encoded by genes which are often found translocated in leukaemia, and whose inactivation has been described in some cancers.[50] Also, mice deficient for Suv39H1, one of the Rb-associated histone H3 K9-specific HMTs, and for the related Suv39H2 develop tumours.[51] Finally, SNF5, a component of the SWI/SNF complex, is deleted in virtually all malignant rhabdoid tumours.[52] It has to be noted, however, that in all these cases, the involvement of the deregulation of the E2F/Rb pathway in tumorogenesis is unclear.

Open Questions

The involvement of some specific proteins acting at the chromatin levels and of some special modifications has been well documented, indicating the existence of "chromatin codes" on E2F-regulated promoters that bring about the proper functional response, such as activation, transitory repression or stable repression. However, this "code" is far from being entirely deciphered. For example, the occurrence of some histone modifications which are known to be important in other organisms (such as H2B ubiquitinylation) or on other promoters (histone phosphorylation) has not been investigated on E2F-regulated promoters. Moreover, some proteins (HATs leading to histone acetylation at the end of G1) or molecular mechanisms (demethylation during G1 progression, see Fig. 4) responsible for setting up this code are not characterized for the moment. Finally, nearly nothing is known about how this code is read, and how that leads to the precise transcriptional response of E2F-regulated promoters.

Acknowledgements

DT's group is supported by the "Ligue Nationale Contre le Cancer", as an "équipe labellisée". LD and EN are respectively recipient of a postdoctoral fellowship from the "Ligue Nationale Contre le Cancer" and a studentship from the "Association pour la Recherche sur le Cancer".

References

1. Weinberg RA. The retinoblastoma protein and cell cycle control. Cell 1995; 81(3):323-30.
2. Mulligan G, Jacks T. The retinoblastoma gene family: Cousins with overlapping interests. Trends Genet 1998; 14(6):223-9.
3. Muller H, Helin K. The E2F transcription factors: Key regulators of cell proliferation. Biochim Biophys Acta 2000; 1470(1):M1-12.
4. de Bruin A, Maiti B, Jakoi L et al. Identification and characterization of E2F7, a novel mammalian E2F family member capable of blocking cellular proliferation. J Biol Chem 2003; 278:42041-9.
5. Trimarchi JM, Fairchild B, Verona R et al. E2F-6, a member of the E2F family that can behave as a transcriptional repressor. Proc Natl Acad Sci USA 1998; 95(6):2850-5.
6. Morkel M, Wenkel J, Bannister AJ et al. An E2F-like repressor of transcription. Nature 1997; 390(6660):567-8.
7. Weintraub SJ, Prater CA, Dean DC. Retinoblastoma protein switches the E2F site from positive to negative element. Nature 1992; 358(6383):259-61.
8. Ross JF, Naar A, Cam H et al. Active repression and E2F inhibition by pRB are biochemically distinguishable. Genes Devel 2001; 15(4):392-7.
9. Strahl BD, Allis CD. The language of covalent histone modifications. Nature 2000; 403(6765):41-5.
10. Eberharter A, Becker PB. Histone acetylation: A switch between repressive and permissive chromatin. EMBO Rep 2002; 3:224-29.
11. Dhalluin C, Carlson JE, Zeng L et al. Structure and ligand of a histone acetyltransferase bromodomain. Nature 1999; 399(6735):491-6.
12. Jacobson RH, Ladurner AG, King DS et al. Structure and function of a human TAFII250 double bromodomain module. Science 2000; 288(5470):1422-5.
13. Bannister AJ, Zegerman P, Partridge JF et al. Selective recognition of methylated lysine 9 on histone H3 by the HP1 chromo domain. Nature 2001; 410:120-24.
14. Lachner M, O'Carrol D, Rea S et al. Methylation of histone H3 lysine 9 creates a binding site for HP1 proteins. Nature 2001; 410:116-20.
15. Jenuwein T, Allis CD. Translating the histone code. Science 2001; 293(5532):1074-80.
16. Legube G, Trouche D. Regulating histone acetyltransferases and deacetylases. EMBO R 2003; 4:944-47.
17. Trouche D, Cook A, Kouzarides T. The CBP coactivator stimulates E2F1/DP1 activity. Nucleic Acids Res 1996; 24(21):4139-45.
18. Bannister AJ, Kouzarides T. The CBP coactivator is a histone acetyltransferase. Nature 1996; 384(6610):641-3.
19. Ogryzko VV, Schiltz RL, Russanova V et al. The transcriptional coactivators p300 and CBP are histone acetyltransferases. Cell 1996; 87(5):953-59.
20. Lang SE, McMahon SB, Cole MD et al. E2F transcriptional activation requires TRRAP and GCN5 cofactors. J Biol Chem 2001; 276(35):32627-34.
21. Magnaghi-Jaulin L, Groisman R, Naguibneva I et al. Retinoblastoma protein represses transcription by recruiting a histone deacetylase. Nature 1998; 391:601-05.
22. Luo RX, Postigo AA, Dean DC. Rb interacts with histone deacetylase to repress transcription. Cell 1998; 92(4):463-73.
23. Brehm A, Miska EA, McCance DJ et al. Retinoblastoma protein recruits histone deacetylase to repress transcription. Nature 1998; 391(6667):597-601.
24. Lai A, Lee JM, Yang WM et al. RBP1 recruits both histone deacetylase-dependent and -independent repression activities to retinoblastoma family proteins. Mol Cell Biol 1999; 19(10):6632-41.
25. Ferreira R, Magnaghi-Jaulin L, Robin P et al. The three members of the pocket proteins family share the ability to repress E2F activity through recruitment of a histone deacetylase. Proc Natl Acad Sci USA 1998; 95(18):10493-8.
26. Ferreira R, Naguibneva I, Mathieu M et al. Cell cycle-dependent recruitment of HDAC-1 correlates with deacetylation of histone H4 on an Rb-E2F target promoter. EMBO Rep 2001; 2(9):794-9.
27. Takahashi Y, Rayman JB, Dynlacht BD. Analysis of promoter binding by the E2F and pRB families in vivo: Distinct E2F proteins mediate activation and repression. Genes Dev 2000; 14(7):804-16.
28. Rayman JB, Takahashi Y, Indjeian VB et al. E2F mediates cell cycle-dependent transcriptional repression in vivo by recruitment of an HDAC1/mSin3B corepressor complex. Genes Devel 2002; 16(8):933-47.

29. Nicolas E, Morales V, Magnaghi-Jaulin L et al. RbAp48 belongs to the histone deacetylase complex that associates with the retinoblastoma protein. J Biol Chem 2000; 275(13):9797-804.
30. Polanowska J, Fabbrizio E, Le Cam L et al. The periodic down regulation of cyclin E gene expression from exit of mitosis to end of G(1) is controlled by a deacetylase- and E2F-associated bipartite repressor element. Oncogene 2001; 20(31):4115-27.
31. Lu X, Horvitz HR. lin-35 and lin-53, two genes that antagonize a C. elegans Ras pathway, encode proteins similar to Rb and its binding protein RbAp48. Cell 1998; 95(7):981-91.
32. Chen Z, Han M. C. elegans Rb, NuRD, and Ras regulate lin-39-mediated cell fusion during vulval fate specification. Current Biology 2001; 11(23):1874-9.
33. Kouzarides T. Acetylation: A regulatory modification to rival phosphorylation? Embo J 2000; 19(6):1176-79.
34. Martinez-Balbas MA, Bauer UM, Nielsen SJ et al. Regulation of E2F1 activity by acetylation. Embo J 2000; 19(4):662-71.
35. Marzio G, Wagener C, Gutierrez MI et al. E2F family members are differentially regulated by reversible acetylation. J Biol Chem 2000; 275(15):10887-92.
36. Chan HM, Krstic-Demonacos M, Smith L et al. Acetylation control of the retinoblastoma tumour-suppressor protein. Nat Cell Biol 2001; 3(7):667-74.
37. Vandel L, Nicolas E, Vaute O et al. Transcriptional repression by the retinoblastoma protein through the recruitment of a histone methyltransferase. Mol Cell Biol 2001; 21(19):6484-94.
38. Nielsen SJ, Schneider R, Bauer UM et al. Rb targets histone H3 methylation and HP1 to promoters. Nature 2001; 412(6846):561-5.
39. Nicolas E, Roumillac C, Trouche D. Balance between acetylation and methylation of histone H3 lysine 9 on the E2F-responsive dihydrofolate reductase promoter. Mol Cell Biol 2003; 23(5):1614-22.
40. Ogawa H, Ishiguro K, Gaubatz S et al. A complex with chromatin modifiers that occupies E2F- and Myc- responsive genes in G0 cells. Science 2002; 296(5570):1132-6.
41. Ghosh MK, Harter ML. A viral mechanism for remodeling chromatin structure in G0 cells. Mol Cell 2003; 12(1):255-60.
42. Narita M, Nunez S, Heard E et al. Rb-mediated heterochromatin formation and silencing of E2F target genes during cellular senescence. Cell 2003; 113(6):703-16.
43. Czermin B, Schotta G, Hulsmann BB et al. Physical and functional association of SU(VAR)3-9 and HDAC1 in Drosophila. EMBO Rep 2001; 2(10):915-9.
44. Vaute O, Nicolas E, Vandel L et al. Functional and physical interaction between the histone methyl transferase Suv39H1 and histone deacetylases. Nucleic Acids Res 2002; 30(2):475-81.
45. Zhang Y, Reinberg D. Transcription regulation by histone methylation: Interplay between different covalent modifications of the core histone tails. Genes Dev 2001; 15(18):2343-60.
46. Trouche D, Le Chalony C, Muchardt C et al. RB and hbrm cooperate to repress the activation functions of E2F1. Proc Natl Acad Sci USA 1997; 94(21):11268-73.
47. Cosma MP, Tanaka T, Nasmyth K. Ordered recruitment of transcription and chromatin remodeling factors to a cell cycle- and developmentally regulated promoter. Cell 1999; 97(3):299-311.
48. Zhang HS, Dean DC. Rb-mediated chromatin structure regulation and transcriptional repression. Oncogene 2001; 20(24):3134-8.
49. Dahiya A, Wong S, Gonzalo S et al. Linking the Rb and polycomb pathways. Mol Cell 2001; 8(3):557-69.
50. Timmermann S, Lehrmann H, Polesskaya A et al. Histone acetylation and disease. Cell Mol Life Sci 2001; 58(5-6):728-36.
51. Peters AH, O'Carroll D, Scherthan H et al. Loss of the Suv39h histone methyltransferases impairs mammalian heterochromatin and genome stability. Cell 2001; 107(3):323-37.
52. Versteege I, Sevenet N, Lange J et al. Truncating mutations of hSNF5/INI1 in aggressive paediatric cancer. Nature 1998; 394(6689):203-6.
53. Bannister AJ, Schneider R, Kouzarides T. Histone methylation: Dynamic or static? Cell 2002; 109(7):801-6.

CHAPTER 8

Emerging Roles for the Retinoblastoma Gene Family

Jacqueline L. Vanderluit, Kerry L. Ferguson and Ruth S. Slack*

Abstract

Research on the retinoblastoma protein has grown from studying its role as a tumour suppressor in cancer to identifying it as a key regulator of the cell cycle G1/S check point and today to exploring its function in numerous cellular processes. The recent development of conditional knockout mice has shed new light on the roles of Rb in embryonic development and has aided in the identification of the cell-of-origin in Retinoblastoma cancer. In this review, we will discuss the role of Rb as a tumour suppressor as well as its role in cell division, differentiation, apoptosis and cancer.

Identification of Rb as a Tumour Suppressor

The retinoblastoma susceptibility gene (*RB*) gene was the first tumour suppressor gene to be identified. It was initially discovered due to its mutation in the rare pediatric eye tumour, retinoblastoma.[76,77,148] Retinoblastoma tumours can occur as sporadic or hereditary cases, and can be used as a paradigm for tumourigenesis through loss-of-function mutations.[245] In familial cases of retinoblastoma, young children develop bilateral multifocal retinal tumours, such that individuals carrying a germline mutation for one *RB* allele have a 95% chance of developing retinoblastoma.[60,79] By statistical analysis of the affected families, Knudson proposed that the children inherited one defective autosomal allele and the second wildtype allele was lost during retinal development, leading to retinoblastoma.[128] The deletion or mutational inactivation of one *RB* allele is, therefore, the rate limiting step for the development of these retinal tumours.[15,76,101,128,147] In hereditary cases, patients have over a 30-fold increased risk of developing a second primary tumour, such as osteosarcoma, melanoma, and brain tumours.[66,175] The *RB* gene, which is located on human chromosome 13q, is in an area of the genome frequently lost in sporadic forms of cancer. Mutations are often associated with tumours in several cell types, including sporadic retinoblastoma, osteosarcoma, small cell lung carcinomas, and cancers of the bladder, kidney, prostate and breast.[14,76,77,91,103,143,231,265]

The *RB* gene product, Rb, was identified as a target of oncoproteins expressed by DNA tumour viruses including the adenovirus E1A protein, the simian virus 40 large T antigen (SV40 Tag),[51] human papillomavirus (HPV) E7 protein,[63,182] and the large T antigen of polyomaviruses.[64] As these viral oncoproteins are capable of immortalizing and transforming various cell types, studies examining their properties demonstrated the importance of Rb as a regulator of cell proliferation. By direct binding, these oncoproteins had the capacity to interfere with the growth suppressive functions of Rb. Inactivation of Rb by E1A was essential to

*Corresponding Author: Ruth S. Slack—Department of Cellular Molecular Medicine, Ottawa Health Research Institute, University of Ottawa, 451 Smyth Road, Ottawa, Ontario, K1H 8M5, Canada. Email: rslack@uottawa.ca

Rb and Tumorigenesis, edited by Maurizio Fanciulli. ©2006 Eurekah.com and Springer Business+Science Media.

drive cells into a proliferative state.[65,253] Mutations in the Rb binding region of any of these viral oncoproteins abrogated their capacity for viral transformation.[65]

Rb is a 110 kDa nuclear phosphoprotein.[103,148,241,264] It is expressed ubiquitously at similar levels in all human and mouse cells examined, with the exception of tumour cells in which the *RB* gene is inactivated by mutation or deletion.[103,148,241,264] Its status as a tumour suppressor was shown by the fact that tumours develop in the absence of Rb in humans, and reintroduction of Rb into these Rb-deficient tumour cells was sufficient to partially block the malignant phenotype.[111] Further, Rb can be inactivated by constitutive hyperphosphorylation in tumours that do not contain mutations in the retinoblastoma gene.[218] Together, these studies linking Rb disruption to tumour formation and demonstrating Rb to be one of the obligatory cellular targets for viral transformation, have provided strong evidence for Rb as an important regulator of cellular proliferation.

Structure and Functional Domains of Rb Family Members

Rb belongs to the Retinoblastoma family of genes which includes p107 and p130. The family shares sequence homology in their A/B pocket, the domain with which they interact with transcription factors and viral oncoproteins- and are thereby termed "pocket proteins".[68,88,154,171] The Rb protein has a globular structure and contains several domains required for its function.[145] The highly conserved domains A and B, separated by a spacer region, interact to form a central "pocket" structure.[35,145] The pocket is disrupted in most naturally-occurring and tumour-derived mutations in retinoblastoma patients.[89,103] It is now known that the pocket domain consists of binding sites so that more than one protein can bind simultaneously (Fig. 1).

Many viral oncoproteins and endogenous Rb-binding proteins contain an LXCXE motif that allows them to bind Rb pocket proteins.[63,145,160,253] While the LXCXE binding site is located within a shallow groove of domain B,[145] domain A is required for the proper conformation of domain B.[125,145] Endogenous Rb binding proteins which contain an LXCXE-like motif include the chromatin remodelling factors HDACs-1 and -2 and BRG1.[18,59,163a,168] However, the LXCXE motif is not required for binding to the pocket domain. In contrast to HDACs-1 and -2, HDAC-3 does not contain an LXCXE sequence and mutation of the LXCXE binding site in Rb does not inhibit HDAC-3 binding.[32,44] Further, although BRG1 contains an LXCXE sequence, this sequence does not appear to be required to bind Rb.[270]

E2Fs lack an LXCXE sequence and instead contain an Rb-binding motif at their carboxyl terminus.[73,92] While truncation analyses have shown the Rb pocket domain to be sufficient for stable interaction with E2F1,[202] more recent studies have revealed that Rb contains two distinct E2F binding sites.[56] The first binding site, located within the pocket region, is necessary for stable association with DNA. The removal of this site inhibits Rb-mediated growth suppression but maintains regulation of E2F1-induced apoptosis. A second site in the carboxy-terminal region of Rb is specific for E2F1. Rb-E2F1 complexes at this site have low affinity for DNA, but are required to regulate E2F1-mediated apoptosis.[56] The distinct Rb binding sites allow E2Fs to bind simultaneously with other proteins, such as those with an LXCXE sequence.[2,56]

The carboxy-terminal region of Rb contains binding sites for the cellular homologue of the transforming sequence of Abelson murine leukemia virus (c-abl) tyrosine kinase and murine double minute 2 (MDM-2), which appear to be distinct from the E2F-binding site.[249,263] The tyrosine kinase function of c-abl is blocked when it is bound to Rb, which appears to be important for Rb-mediated growth suppression.[249,252]

In contrast, the amino-terminal region of Rb seems to be dispensable for growth repression.[202,266,267] Overexpression of a truncated Rb mutant which lacked the amino-terminus, retained the ability to suppress the proliferation of tumour cells.[266] While not necessary for growth regulation, this region may be important for other functions of Rb. Several proteins interact with the Rb amino-terminus including the transcription factor Sp1, minichromosome

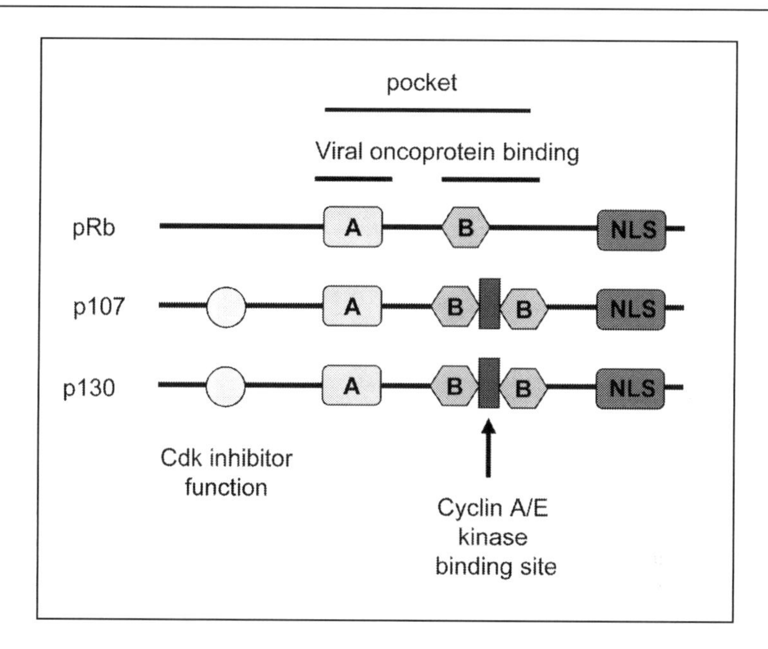

Figure 1. The Retinoblastoma gene family. Rb, p107 and p130 share the highly conserved domains A and B of the Rb protein and are separated by a spacer region to form a central "pocket" structure. Rb family members interact with the majority of proteins, including those containing an LXCXE sequence, through the pocket domain. E2F binding requires both the central pocket region and the carboxy terminus.

maintenance 7 (MCM7), a replication licensing factor, and Rb/histone H1 kinase (RbK) which is a G2/M cycle-regulated kinase.[227,238] The region also contains several consensus cyclin dependent kinase (CDK) phosphorylation sites, which may be important for Rb cell-cycle regulation. The amino-terminus has been suggested to be important for Rb function developmentally by studies in which the embryonic lethality of Rb knockouts was delayed by reintroduction of an Rb transgene with an amino-terminus mutation.[206]

Rb family members share a high sequence homology across the pocket domain as well as in their C-terminus including the nuclear localizing signal (NLS). p107 and p130 however share a closer homology with each other than with Rb. For instance, both p107 and 130 have a high affinity binding site for cyclin A/CDK2 and cyclin E/CDK2 inserted in their spacer region between the A and B domains of their respective pockets which is absent in Rb (Fig. 1).[69,70,149] The amino terminal domains of p107 and p130 contain a region of homology that appears to function as a CDK inhibitor and their B domains contain a spacer insertion, function currently unknown[26,257] for review see ref. 37. Hence the homology across Rb pocket proteins allows for partial compensation in the absence of a family member, whereas their differences provide for their distinct regulation and activity.

The Rb Family Regulates the Cell Cycle

The importance of Rb function in tumour suppression and with viral oncoproteins suggested that Rb proteins may have a role in regulating normal cellular proliferation. The mammalian cell cycle is divided into four distinct phases, referred to as G1, S, G2, and M phases and as cells exit the cell cycle to differentiate or to become quiescent they enter G0. The two gap periods, G1 and G2, are growth phases, which are followed, respectively, by DNA synthesis and replication (S phase) and mitosis (M phase).[217] Rb family members are differentially expressed throughout the cell cycle indicative of distinct roles. Rb is expressed at moderate levels throughout the cell cycle and in differentiated cells in G0, but its expression is highest

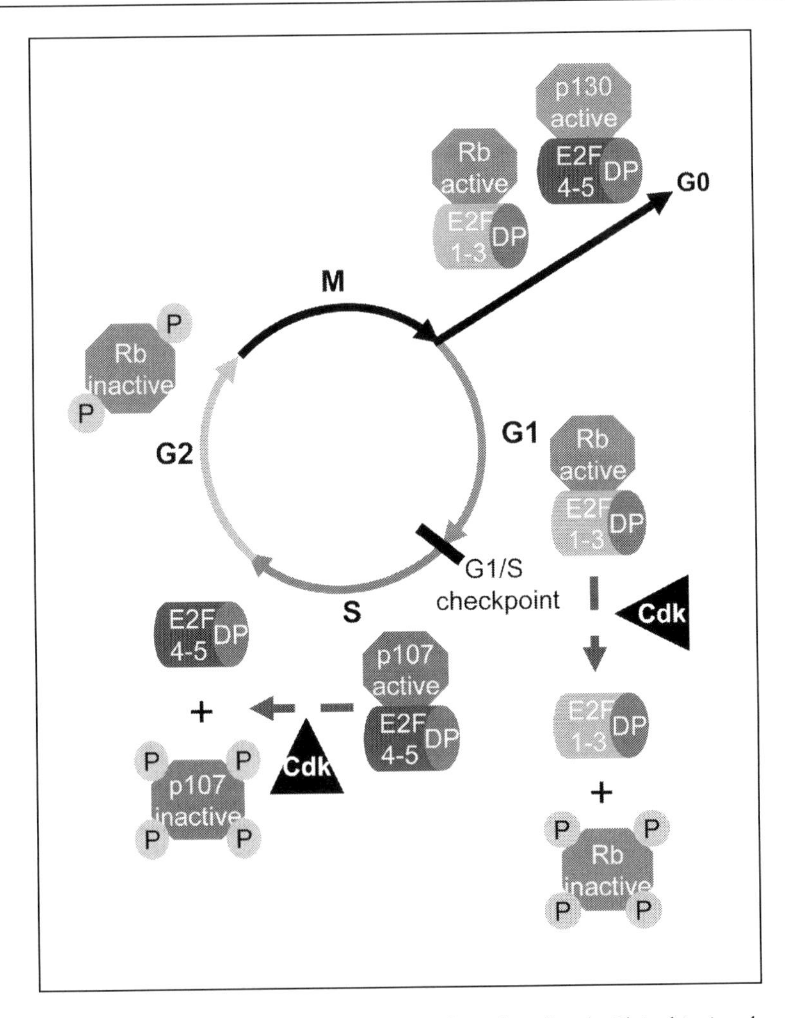

Figure 2. Expression pattern of Rb pocket proteins throughout the cell cycle. Rb is ubiquitously expressed in cycling and differentiated cells, however its expression peaks as cells enter the cell cycle in G1. P107 is expressed in cycling cells and being an E2F responsive gene its expression is upregulated in G1 and S phase of the cell cycle as Rb mediated E2F repression is lifted. P130 expression is undetectable in cycling cells, its expression increases as cells exit the cell cycle in G0 and can be found ubiquitously expressed in post-mitotic cells.

during G1. P107 expression fluctuates during the cell cycle, increaseing as cells enter the cell cycle in G1 and throughout S-phase and rapidly down-regulated in post-mitotic cells.[9,98,124] The expression of p130, sharply contrasts Rb and p107, as it is not expressed during the cell cycle but is specifically induced as cells exit the cell cycle in G0.[41,225] Hence, Rb has been shown to have a primary role in G1/S progression. P107 controls the cell cycle to a lesser extent whereas p130 appears to have more of a role in differentiated cells.

In late G1, cells pass through the restriction point, at which they commit to complete the cell cycle, independently of further growth factor signalling. Passage through the restriction point and entry into S phase is regulated by Rb phosphorylation by CDKs.[31,99,162] When underphosphorylated, Rb is active and able to bind to and repress transcription factors which promote proliferation, most notably, the E2F family of transcription factors. Phosphorylation

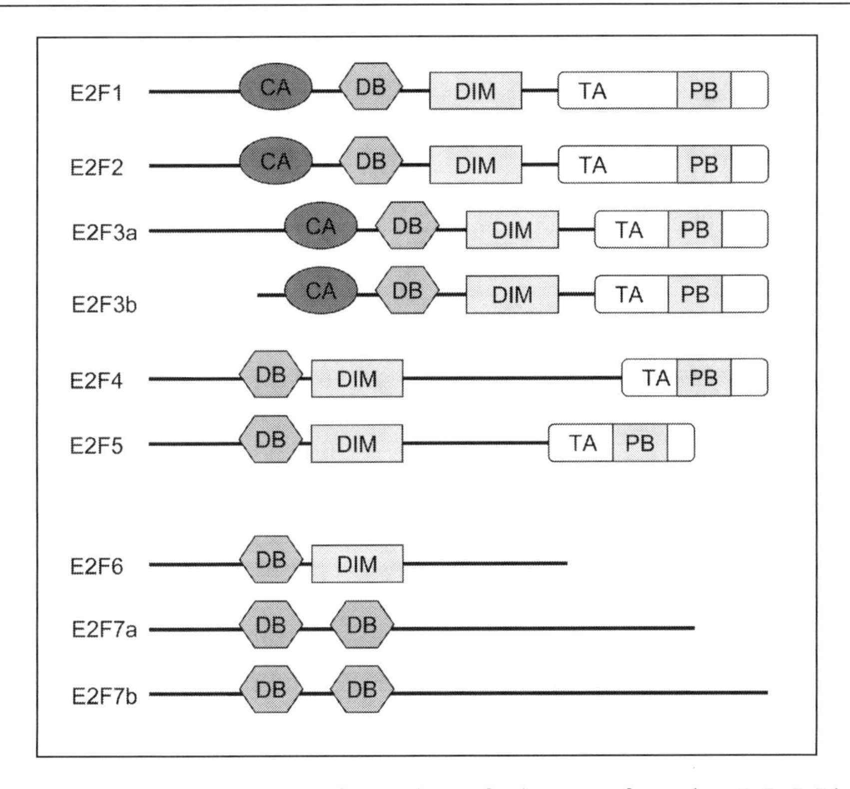

Figure 3. The E2F family of transcription factors. The E2F family consists of 7 members. E2F1-E2F6 are characterized by their shared DIM domain for dimerization with DP and their DB domain for DNA binding. In contrast to E2F1-6, E2F7a and E2F7b do not interact with DP transcription factors. E2F1-E2F5 also share a PB domain for binding pocket proteins and a TA domain for transcriptional activation. E2F1-E2F3b also bind cyclin A through a CA domain in their N-terminus.

by CDKs inactivates Rb pocket proteins, thereby releasing these bound transcription factors and driving S phase progression[22] (Fig. 2). Overexpression of Rb induced cells to arrest in G1 of the cell cycle,[85,111,202] whereas cells deficient for Rb had an accelerated G1-S phase transition.[39,95,112]

In late G1 and early S-phase while Rb is becoming phosphorylated, p107 is underphosphorylated and able to bind and repress a different set of E2F transcription factors to regulate S-phase progression (Fig. 2). In the absence of p107, cells exhibit a more rapid S-phase progression.[141] Overexpression of p107 arrests cells in G1 and unlike Rb, this arrest is dependent on both the pocket domain and the spacer region.[224,225]

Rb Family Members Interact with the E2F Family of Transcription Factors

The first identified cellular target of Rb was the E2F1 transcription factor.[8,28,94,122,216] E2F1 was first identified as a cellular factor required for the adenovirus early region 1A (E1A)-transforming protein to mediate the transcriptional activity of the viral E2A promoter.[187] The importance of E2F1 in cell proliferation was shown by studies in which E2F1 overexpression was found to be sufficient to promote the G1-S phase transition.[120] Further, Rb was able to arrest cells in G1 by inhibiting E2F1 transactivation.[204,276] Since then, a whole family of E2F transcription factors has been identified, most of which interact with Rb or its closely related family members, p107 and p130.

In mammalian cells, seven E2F family members have now been identified (E2F1-E2F7) and three of their obligate binding partners (DP1-DP3).[7,8,23,28,34,47,55,94,135,273] (Fig. 3). E2F transcriptional activity arises from the formation of a heterodimer with one member of the E2F family bound to one member of the DP family. All possible combinations of E2F/DP complexes can exist in vitro, allowing the potential for a wide array of cellular E2F complexes. E2F consensus binding sites are present in many genes including cell cycle regulators such as cdc2, *myc*, *b-myb* and cyclins D1, A and E; enzymes required for DNA synthesis such as dihydrofolate reductase (DHFR), thymidine kinase, and DNA polymerase α; the pocket proteins Rb and p107; and E2F1 itself.[16,45,52,67,87,97,123,136,180,186,215,220,225,233,274] In vitro, different E2F/DP complexes recognize similar nucleotide sequences,[23,150,273] however, there is also evidence that some heterodimers preferentially bind to specific E2F sequences.[232] It is not yet clear whether specific E2F complexes target preferred E2F-regulated promoters.

The E2F family members can be categorized into four groups based on sequence homology, with functional similarities apparent in each subgroup. For instance, the expression of E2Fs 1-3, is cell-cycle regulated and peaks in late G1,[107,119,150,151,185,212] while E2Fs 4 and 5 are more uniformly expressed over the cell cycle. The preference with which E2F/DP heterodimers associate with the pocket proteins appears to be specified by the E2F family member: E2Fs 1-3 preferentially bind Rb, whereas E2Fs 4 and 5 interact predominantly with p107 and p130[10,82,98,150,204] reviewed in refs. 228, 236. In contrast, E2Fs 6 and 7 which compose the last two subgroups respectively, lack the pocket protein binding domain and do not interact with Rb family members and will not be further discussed here[47,55,177] for a recent review see.[17]

E2Fs 1-3, which represent the first subgroup, activate E2F-responsive genes and drive cellular proliferation.[54] Overexpression of each one is sufficient to induce quiescent cells to reenter the cell cycle,[6,120,161,244] and expression of a dominant-negative DP mutant has been shown to block S phase entry.[259] The combined deletion of E2Fs 1, 2 and 3 results in a complete blockage of S phase entry.[260] Expression of these E2F factors is sufficient to override growth suppression by various CKIs and cell cycle arrest induced by a dominant-negative CDK2.[52,161,169] This ability is dependent upon the dimerization, DNA-binding, and transactivation domains, suggesting that transcription of E2F-regulated genes is required for S phase entry.[120,203]

The E2F3 locus encodes two protein products: E2F3a, and a newly identified transcript, E2F3b, which is encoded by a unique mRNA transcribed from an intronic promoter.[1,152] E2F3a is cell cycle regulated like E2Fs 1 and 2, whereas, E2F3b is constitutively expressed throughout the cell cycle similar to E2Fs 4 and 5. E2F3b has been shown to maintain preferential association with Rb in G0 cells which may reflect the previously described role of Rb as a transcriptional repressor in quiescent cells.[1,152]

The second E2F subgroup, E2Fs 4 and 5, are believed to act as repressors of E2F responsive genes. Expression of these E2Fs has been shown to induce cell cycle withdrawal and differentiation in various cell types, including adipocytes, keratinocytes, and neural precursors.[57,71,137,156,196,199] Their activity may, in part, be due to their subcellular localization. They are predominantly cytoplasmic and require assembly either with a pocket protein or with DP2, which unlike DP1 contains an NLS, for nuclear import.[3,49,155,167,242] However, significant levels of E2Fs 4 and 5 are present in the nucleus of quiescent (G0) cells.[81] During G0/G1, significant levels of E2F4/p130 complexes occupy E2F sites, with low levels of local histone acetylation.[230,239] Reduced E2F4 occupancy occurs in late G1, correlating with the timing of gene induction, the appearance of E2Fs 1-3, and increased acetylation of histones H3 and H4. The reduced E2F4 promoter occupancy also occurs at the same time as dissociation of E2F4/p130 complexes and relocation of E2F4 to the cytoplasm.[230] Export of E2F4 from the nucleus has been demonstrated to be an active process, and to prevent its ability to induce cell cycle arrest.[81] The fact that E2F4- and 5-pocket protein complexes predominate in G0/G1 cells suggests that they may be important for repression of early cell cycle progression.[174] Consistent with this, mutation of the E2F binding site in certain E2F-responsive promoters, including B-*myb*, cdc2, and E2F1, results in increased expression during G0/G1.[45,107,136]

Mechanisms of Rb-Mediated Transcriptional Repression

It is now widely believed that Rb family members regulate cell cycle progression and differentiation through repression of E2F-dependent gene transactivation. Repression can occur in two ways: First, Rb members bind transcription factors such as E2F and repress their ability to activate transcription.[73,92] Since the pocket protein binding domain of E2F is integrated within its transactivation domain, Rb members block E2F activity by binding and thereby masking the E2F transactivation domain (Fig. 3).[73,92]

Second, Rb family members can actively repress transcription by recruitment of chromatin remodelling enzymes to the Rb-E2F repressor complex.[20,213,248] Once targeted to the promoter by E2F, Rb recruits proteins involved in chromatin modification such as histone deacetylases (HDACs), and ATPases of the SWI/SNF complex, brahma (BRM) and brahma-related gene 1 (BRG1).[59,223] The HDACs remove acetyl groups from lysine residues within the tails of histone octamers, which facilitates the condensation of nucleosomes into inactive chromatin.[90] The chromatin condensation, in turn, prevents access of transcription factors to the promoter.[126,130,256]

Rb has also been shown to repress transcription through recruitment of the Rb binding protein, RBP1.[133,134] RBP1 acts as a corepressor, capable of repressing E2F-mediated transcription via its association with the Rb pocket.[133] RBP1 also acts as a bridging molecule by recruiting HDACs with one repression domain and through a second domain, repressing in an HDAC-independent manner.[134]

Inactivation of Rb Family Members by CDK-Dependent Phosphorylation

The variable phosphorylation state of the pocket proteins during the cell cycle acts to regulate its activity. Quiescent cells, or cells in early G1, have mainly hypophosphorylated Rb. As cells progress through late G1 and into S phase, Rb becomes increasingly phosphorylated, and remains this way until it becomes dephosphorylated in late stages of mitosis. Rb is in a hypophosphorylated state in cells exiting the cell cycle to undergo senescence or terminal differentiation[22,31,50,173] (Fig. 2). Rb has 16 possible phosphorylation sites. Each cyclin-CDK complex phosphorylates Rb on particular phosphoacceptor sites.[43,127] Many, if not all of these sites, must be phosphorylated in a sequential manner by different cyclin-CDK complexes, in order to inactivate Rb, reviewed in ref. 2.

In early to mid G1, mitogen stimulation induces the synthesis of D cyclins (D1, D2, and D3). These D cyclins then assemble with their catalytic partners to form cyclin D-CDK4/6 complexes, which phosphorylate Rb on distinct sites. Rb becomes further phosphorylated by cyclin E-CDK2 complexes in late G1.[58,129,191] The cyclin E gene is E2F responsive so cyclin E-CDK2 complexes act in a positive feedback loop to facilitate progressive Rb phosphorylation and E2F release.[162] This feedback loop produces a rapid rise in cyclin E-CDK2 needed to allow cells to enter S phase. Cyclin A-CDK2 complexes become activated at the G1/S phase boundary and throughout S phase.[84,194,278] During G2/M, specific phosphatases remove the inactivating phosphate groups from Rb, enabling the binding of E2Fs before the cells enter a new cycle.[27,159]

As Rb mediated repression is lifted in late G1 and early S-phase, E2F complexes activate E2F responsive genes, including p107. Expression of p107 increases in late G1 and throughout S phase of the cell cycle.[9] P107 interacts with E2Fs 4 and 5 in its active unphosphorylated state. Similar to Rb, as cells progress through S phase p107 becomes progressively phosphorylated by cyclin D dependent kinases and releases E2Fs 4 and 5.[9,262] Unlike pRb however, p107 can also bind CDKs on its amino-terminal and appears to act as a CDK inhibitor.[26,257]

Rb Activation by CDK Inhibitors

CDK activity is negatively regulated by two families of CDK inhibitors (CKIs). The inhibitors of CDK4 (INK4) family which is comprised of p16^{INK4a}, p15^{INK4b}, p18^{INK4c}, and p19^{INK4d}, specifically inhibit the cyclin D-associated kinases, CDK4 and CDK6.[208] The second family,

Cip/Kip, include p27^{Kip1}, p21^{Cip1}, and p57^{Kip2}. The Cip/Kip inhibitors act more generally, and affect the activities of cyclin D-, cyclin E- and cyclin A-dependent kinases.[219] Both p27^{Kip1} and p21^{Cip1} inhibit cyclin E-CDK2 complexes, but are less effective at blocking the enzymatic activity of cyclin D-CDK4.[13,226] In contrast, physical association with Cip/Kip subunits has been shown to facilitate the assembly, stability, and nuclear retention of cyclin D1-CDK4 complexes.[4,33,132,183]

The Overlapping and Distinct Roles of Rb Family Members

All three Rb family members, Rb, p107 and p130 are involved in regulation of the cell cycle.[40,276] Each member facilitates growth arrest in response to expression of the CKI, p16^{INK4a}.[21] Rb itself, however, does not appear to be essential for cell cycle control.[114] Recruitment of HDACs to E2F-responsive promoters in normally cycling fibroblasts is performed mainly by p107 and p130.[205] It has been proposed that Rb may contribute to gene repression only at certain times, for instance, during differentiation or senescence.[157,214,234]

Rb family members share many overlapping roles and have a partially compensatory ability. In the nervous system, Rb is expressed in cycling and differentiated cells, whereas p107 is present only in proliferating cells, becoming rapidly down-regulated upon differentiation.[12,25,117,240] In contrast, p130 is expressed in post-mitotic cells. In Rb deficient embryos, p107 expression is up-regulated and can partially compensate for the loss of Rb.[25] However, the absence of p107 from differentiating and post-mitotic neurons may explain the partial rescue.

While Rb deficiency results in embryonic lethality at mid-gestation,[36,115,144] mice lacking p107 or p130 develop normally on most genetic backgrounds.[42,140,141,146,210] However, on a BalbC genetic background, p107 deficiency results in growth retardation and myeloid hyperplasia,[141] and p130 loss induces embryonic lethality.[140] Compound mutations such as Rb/p107 and Rb/p130 double null mice die earlier during embryogenesis and exhibit more pronounced cell cycle defects and apoptosis.[146,210] Mice deficient for p107/p130 die shortly after birth and display defective bone development.[42] The more severe phenotype of these compound mutant mice implies a partial redundancy in the function of these Rb family members.

Rb members can also have distinct, nonredundant functions. For example, they have specific binding preferences for the various E2Fs: Rb complexes primarily with E2Fs 1-3, although interactions with E2F4 have been detected; while p107 and p130 associate mainly with E2Fs 4 and 5.[98,150,174,239,258] Binding interactions also depend on the phase of cell cycle. Rb associates with E2Fs in both cycling and quiescent cells, but p130-E2F complexes form mostly in G0, while p107 predominantly binds E2Fs during late G1 and S phase.[41,221] In addition, both p107 and p130 but not pRb interact with cyclinE/CDK2 and cyclinA/CDK2.[69,70,149,154] While cells isolated from p107- or p130-deficient embryos have a shortened G1 phase, similar to Rb,[38,112] the genes that are deregulated due to loss of p107 or p130 differ from those associated with Rb deficiency.[112,181,230,251] MEFs deficient for all three Rb family members had a shorter cell cycle compared to wild type cells, were insensitive to various G1 arrest signals, and became directly immortalized.[46,210]

The roles of each Rb family member also appear to depend on the tissue type examined. For instance, in an assay of adipocyte differentiation, Rb-deficient fibroblasts failed to differentiate in response to a cocktail of adipogenesis-inducing agents; whereas p107/p130-deficient cells underwent adipocytic differentiation at a much higher frequency than wildtype cells.[38] In the brain, p107 has been found to regulate the number of stem cells in both embryonic and adult mice, such that p107-null mice have greatly elevated stem cell numbers. In contrast, Rb deficiency has no effect on stem cell proliferation, in spite of being an important regulator of neuronal development.[240] Further, in cerebellar development, the requirements for Rb and p107 depend on the cell type examined.[170] The regulation of cell cycle exit, differentiation, and survival of granule cell precursors was found to be Rb dependent, and p107 was unable to fully compensate for its deficiency. In contrast, neither Rb nor p107 was required for the differentiation and survival of Purkinje neurons.[170] Taken together, these studies demonstrate the essential role of these proteins in the control of the G1-S transition and highlight the shared versus distinct roles in the control of cellular processes, such as differentiation.

The Roles of Rb Family Members in the Developing Embryo

The unique and overlapping roles of Rb family members are perhaps best observed during embryonic development. In situ hybridization experiments reveal distinct spatial and temporal expression patterns of Rb family members. Rb expression is first detected at E8.5 in the nervous system and by E10.5 is highly expressed in the hematopoeitic system as well as the liver, muscle, lens and the retinal ganglion cell layer of the eye.[117,157,268] P107 expression overlaps with Rb in the liver and CNS but its onset of expression is first detected at E10.5. By E14.5, p107 is widely expressed throughout the developing embryo including the lungs, heart, kidney, intestine and cartilage.[117] In contrast to Rb and p107, p130 is expressed late in embryonic development at E15.5 in the developing bones and liver.[117] The overlapping expression profiles of p107 and p130 allow for partial compensation by co-expressed family members. For instance, individual p107-/- and p130-/- mice survive with minimal defects, however double knockout p107-/-:p130-/- mice die at birth with major deficits in bone development.[42,146]

The role of Rb family members in development has historically been based on the phenotype of the null mutants. The pleiotropic defects observed in the Rb mutant however, has led to significant confusion of the real role of Rb in development. Only with the recent generation of chimeric and conditional Rb mutants has the role of Rb been clarified.

Characterization of the Rb Null Phenotype

Germline deletion of Rb in mice is embryonic lethal. Rb-/- die by mid-gestation, between embryonic days 12-15 with defects in erythroid, lens, skeletal muscle and neuronal development.[36,115,144] Specifically, Rb-/- embryos have smaller livers, containing enlarged sinusoids and reduced hepatocytes. In addition, there are increased numbers of immature, nucleated red blood cells in the liver and peripheral circulation.[36,115,144] Increased apoptosis was observed in skeletal muscle precursor cells and Rb-/- cells died prior to the completion of myogenesis.[118,139] In the lens and central and peripheral nervous systems Rb deficiency resulted in widespread apoptosis and ectopic mitoses.[36,115,142,144,176] Lens defects included disorganized histological appearance, cataracts, and defective lens fibre differentiation and elongation.[176] In the nervous system, proliferating neural precursor cells in the ventricular zones appeared normal. However, within the intermediate zones of the developing neural tube, through which newly committed post-mitotic neurons migrate and which do not normally contain dividing cells, there was an abundance of abnormal mitoses and apoptotic death.[142] Expression of a number of markers of neuronal differentiation, including the neuron-specific βIII tubulin, and the neurotrophin receptors TrkA, TrkB, and p75, were all significantly decreased, particularly in dorsal root ganglia.[142]

The Role of Rb in Development as Revealed by Conditional Knockouts

The development of conditional Rb-/- mice clarified the specific role of Rb in embryonic development of the liver, erythrocytes, skeletal muscle, lens and the central and peripheral nervous systems. Rb chimeric mice, in contrast to germline knockouts, survived and exhibited minimal apoptosis.[158,163,255] The survival of Rb chimeric and conditional mice allowed examination of the role of Rb in tissues and cells that develop late in embryogenesis.

The role for Rb in hematopoeisis was assessed in chimeric mice and in transplantation studies. Chimeric analyses, as well as transplantation of Rb-/- cells into irradiated recipients, demonstrated that Rb deficient cells can contribute to all hematopoietic lineages.[109,132,163,255] However, defective erythropoeisis persisted in the wildtype mice transplanted with Rb deficient liver cells.[109] The reconstituted mice were anaemic and showed increased levels of nucleated red blood cells for up to 6 months. It was, therefore, suggested that the defect in Rb null mice may arise from a defective cell in the hematopoietic lineage and not from defective hepatocyte function.[109]

The requirement for Rb in muscle development became evident by the examination of transgenic mice expressing low levels of Rb, driven by an *RB* minigene.[269] The minigene, which consisted of a genomic fragment spanning 1.3kb of the mouse *RB* promoter, with the first exon and intron fused to exons 2 to 27 of the mouse *RB* cDNA, was expressed in Rb-deficient embryos. These mutants survived to birth but exhibited specific skeletal muscle defects, including ectopic proliferation within the myotubes, elevated apoptosis prior to myoblast fusion, shorter myotubes with fewer myofibrils, reduced numbers of muscle fibres, and reduction in expression of the late muscle-specific genes creatine kinase M (MCK) and muscle regulatory factor 4 (MRF4). These results implicated a role for Rb in cell survival and in permanent withdrawal from the cell cycle.[269]

The generation of a conditional knockout in which Rb was specifically deleted in the developing telencephalon revealed the role of Rb in neural development.[72] In spite of virtually complete Rb excision in the forebrain, these mutants survived to birth and exhibited minimal apoptosis. Immunostaining with a mitotic marker demonstrated that precursors did not arrest prior to mitosis, but underwent complete cell divisions outside the normal proliferative regions. Colabeling with the early neuronal marker, TuJ1, revealed that ectopically cycling cells had initiated neuronal differentiation. At E16.5, mutant telencephalic lobes were significantly enlarged, and in some cases, mutants developed cortical tissue protrusions. The enhanced neurogenesis observed in these conditional mutants demonstrated that Rb deficiency in the telencephalon was compatible with neuronal survival and differentiation.[72] This work has subsequently been corroborated by a study in which Rb was deleted from nestin-positive neural precursors.[165] These mutants exhibited ectopic division and enlarged brain size in the absence of increased apoptosis, consistent with enhanced neurogenesis.[165] Hence the ectopic mitosis and increased neurogenesis characteristic of these conditional mutants suggests Rb regulates cell cycle exit and terminal mitosis.

In Rb chimeric mice, ectopic proliferation and extensive cell degeneration were observed in the embryonic retina and the development of lens fibres remained defective.[163,255] The lens defects also failed to be rescued by introduction of a hypomorphic *RB* minigene.[163,255,269] Three recent studies using conditional Rb mutants revealed that selective deletion of Rb in the retina resulted in the death of bipolar and photoreceptor cells.[29,166,272] Together these results demonstrated a cell-specific requirement for Rb for survival.

Another study that shed considerable light on the role of Rb in development came from the lab of Gustavo Leone. An examination of Rb deficiency in the developing extra-embryonic tissue revealed excessive proliferation of trophoblast stem cells which compose the labyrinth layer of the placenta and is the region of nutrient and oxygen exchange between maternal and fetal blood.[261] A significant reduction in essential fatty acids in the embryo at E13.5 was indicative of a malfunctioning placenta. These findings led Leone and colleagues to hypothesize that the embryonic lethality of the germline Rb null embryo was a result of defective placenta development.[261] They subsequently generated mice in which Rb was conditionally deleted from the entire embryo, but were supported by wildtype extra-embryonic tissues, including the placenta.[48,261] These mutants survived to term and did not exhibit massive cell death as was found in germline knockouts. Although the proliferation and apoptosis defects in the lens fibres were not rescued, apoptosis in tissues such as liver and CNS was similar to control levels. In the CNS, mutants exhibited similar proliferation to germline Rb knockouts, including ectopic division.[48,261] The skeletal muscle defects were not rescued and mutants died at birth due to severe muscle dysplasia and hence an inability to respire.[48] These results together with the previous chimeric and conditional Rb mutant studies demonstrate the cell autonomous requirement for Rb in cell cycle regulation, and provide strong evidence that Rb deficiency, and the associated cell cycle perturbations, do not initiate a default apoptotic pathway.

Is There a Role for Rb in the Regulation of Apoptosis?

The lack of an apoptotic phenotype in chimeric and conditional Rb mutants dramatically suggested that the extensive apoptosis observed in the germline Rb knockout was not due to a cell autonomous loss of Rb.[158,163,255] Further, the rescue of the null phenotype with the generation of an Rb null embryo with Rb expressed only in the extra-embryonic tissue revealed that the extensive apoptosis was a result of defective placental development.[48,261] However, in contrast to the liver and CNS, Rb expression in extra-embryonic tissues did not rescue the excessive proliferation and degeneration in the lens fibre cells.[163,255,261] Similarly, chimeric analyses or introduction of a hypomorphic *RB* minigene failed to correct the lens defect.[163,255,269] In the absence of Rb, a cell-specific susceptibility to apoptosis has been further demonstrated in the cerebellum and retina.[29,166,170,272] Conditional Rb deletion in the cerebellum resulted in apoptosis of cerebellar granule cells while Purkinje neurons survived, although they exhibited abnormal cellular morphology.[170] Recently, three independent studies using either Pax6-Cre, Chx10-Cre, or Nestin-Cre demonstrated that selective Rb deficiency in the retina caused apoptosis of photoreceptor and bipolar cells.[29,166,272] These results together imply a cell-specific requirement of Rb for survival.

Cell death as a result of Rb deficiency in the lens and CNS has been shown to involve both p53 and E2F1-dependent pathways.[157] Protein levels and DNA binding activity of p53 are increased in the brains of Rb knockout embryos, along with up-regulation of the p53 target, p21^{Cip1}.[164] Apoptosis in the CNS and lens is rescued in Rb/p53-null animals.[164] Similarly, deletion of E2F1 in Rb-null mice greatly reduced both ectopic mitoses and apoptosis in the lens and CNS, concomitant with a down-regulation of the p53 pathway.[195,209,237,277] These observations led to the conclusion that deregulated E2F1 activity was necessary and sufficient for the p53-mediated apoptosis in Rb deficient embryos.[237]

Several lines of evidence suggest that the release of free E2F1 may be responsible for triggering apoptosis. E2F1 overexpression has been shown to be capable of inducing apoptosis,[54,108,131,190] and E2F1-deficient neurons were protected from certain apoptotic stimuli.[83,104,190] These abilities of E2F1 to drive cell cycle progression and to induce apoptosis can be mechanistically dissociated, since mutants that fail to induce S phase are still capable of inducing apoptosis.[200] Mutational analysis has revealed that the ability of E2F1 to induce apoptosis requires DNA binding but may also function independent of its transactivation domain.[108] However, E2F1 has been shown to directly transactivate several proapoptotic genes including Apafl, caspases-3 and -7, Siva and the BH3-only members of the Bcl2 family, Puma, Noxa, Bim and Hrk/DP5.[74,96,178] In addition, disruption of pro-apoptotic genes such as Apafl and caspase-3, which have been shown to be regulated by E2F1, can partially rescue defects associated with Rb deficiency.[86,222] While E2F1 clearly has a role in mediating apoptosis associated with Rb deficiency, the exact nature of this function remains unclear.

It has been suggested that Rb function in apoptosis may occur through CDK activity. Cyclin D1 transcripts and CDK4/cdc2 protein levels are increased coincident with the death of sympathetic neurons and neuronal PC12 cells following NGF withdrawal.[75,80] Protein levels of cyclin D1, cyclin B, and CDK4 are elevated in brains of Alzheimer's disease and stroke patients.[24,153,172] Stroke injury and DNA damaging agents induced phosphorylation of Rb and p107, followed by loss of Rb and p107.[193,197,198] The CDK inhibitor, flavopiridol, led to neuroprotection, as well as suppression of Rb and p107 phosphorylation and loss.[193,197,198] These studies demonstrate the involvement of CDKs in neuronal cell death, through their ability to phosphorylate Rb.

Rb and Terminal Differentiation

Many of the defects associated with Rb deficiency during development are associated with failed differentiation. While it appears that the majority of cells which are induced to divide inappropriately subsequently undergo apoptosis, there has also been strong evidence of failed differentiation in the surviving cells. Cells of the erythroid, neural, and lens lineages are able to initiate differentiation; however, they often fail to achieve a fully differentiated state. For example, erythrocytes exhibit inefficient enucleation, and some cells of the lens and CNS have reduced or absent expression of specific late differentiation markers.[142,176] In chimeric mice, Rb-deficient cells did not contribute to the post-natal retina.[163,207] Rb deficient cells in both the outer- and the inner-nuclear layers of the retina underwent apoptosis, indicating that Rb may be required at specific stages of retinal differentiation.[29,163,166,207,272] Rb deficiency is also associated with defective myogenesis. Rb deficient fibroblasts induced to differentiate by MyoD are impaired in the acquisition of late differentiation markers and are unable to maintain the differentiated state.[211,189] In addition, Rb-/- fibroblasts exhibit defective adipocyte differentiation, which is related to a direct interaction between Rb and the CCAAT/enhancer binding protein (C/EBP).[30]

While Rb mediated cell cycle arrest has been well studied, its role in differentiation is not as clear. Transcription factors which have been shown to interact with Rb and induce differentiation include c-abl, C/EBP, and the basic helix-loop-helix (bHLH) transcription factor, myogenic determination (MyoD)[30,59,229,250] reviewed in ref. 179.

The bHLH transcription factors have been shown to have important roles in cell fate determination and differentiation in several systems including myogenesis,[247] haematopoiesis[201] and neurogenesis.[116] Inhibitor of differentiation (Id) proteins are members of the HLH family and are important negative regulators of differentiation in virtually all tissues.[188] Since they lack the basic DNA-binding domain, they act as dominant inhibitors of bHLH factors. Through binding and sequestration of E proteins, the required cofactors for bHLH dimerization and activation, Id proteins inhibit bHLH function.[11,188] Of the Id family, Id2 has the additional ability to bind Rb family members, and does so exclusively in their active, hypophosphorylated state.[113,138] By binding to the pocket domain, Id2 abolishes the growth inhibitory functions of Rb.[113,138] Expression of Id2 in cortical progenitor cells was shown to inhibit the induction of neuronal-specific genes, while suppression was eliminated by coexpression of a constitutively active Rb.[235] Many of the defects associated with Rb deficiency have been shown to be rescued by the additional deletion of Id2.[139] In contrast to mid-gestation lethality in Rb-deficient embryos, the Rb/Id2 mutants survived to birth and the haematopoeitic and neurological defects were rescued.[139] Throughout the CNS, there was no evidence of enhanced apoptosis or inappropriate proliferation, including ectopic mitoses.[139] However, the Rb/Id2 mutant mice were born with a severe reduction in muscle mass, as Id2 deficiency was unable to rescue apoptosis in muscle cells.[139] While Id2 appears to have an important role in mediating the Rb deficient phenotype, the mechanism of action remains unknown and may differ greatly depending on the tissue examined.

Rb Family Proteins Interact with the Notch1-Hes1 Signaling Pathway

Recent studies in our laboratory have demonstrated that the Rb family member, p107 interacts with the Notch signaling pathway. The expression of p107 specifically in proliferating cells along the ventricle in both embryonic and adult mice led us to question its role in neural precursor cells. An examination of p107 null mice revealed increased numbers of proliferating progenitor cells in the ventricular subependyma of adult mice.[240] In vitro assays to quantify the number of neural stem cells demonstrated that embryonic and adult p107-/- brains contained higher numbers of stem cells and these stem cells had an enhanced capacity for self-renewal.[240] In contrast, the number of neural stem cells in embryonic Rb -/- was no different than wild type littermate controls implying Rb does not regulate the stem cell pool.

The Notch-Hes signaling pathway is required to maintain the stem cell populations. Previous studies have demonstrated that the Notch signaling pathway is necessary for self-renewing stem cell divisions.[100,184,192] P107-/- neural precursor cells expressed higher levels of Notch1 mRNA, activated Notch1 protein and its downstream target, Hes1 suggesting greater activation of the Notch1 signaling pathway in these cells. Identification of putative E2F binding sites

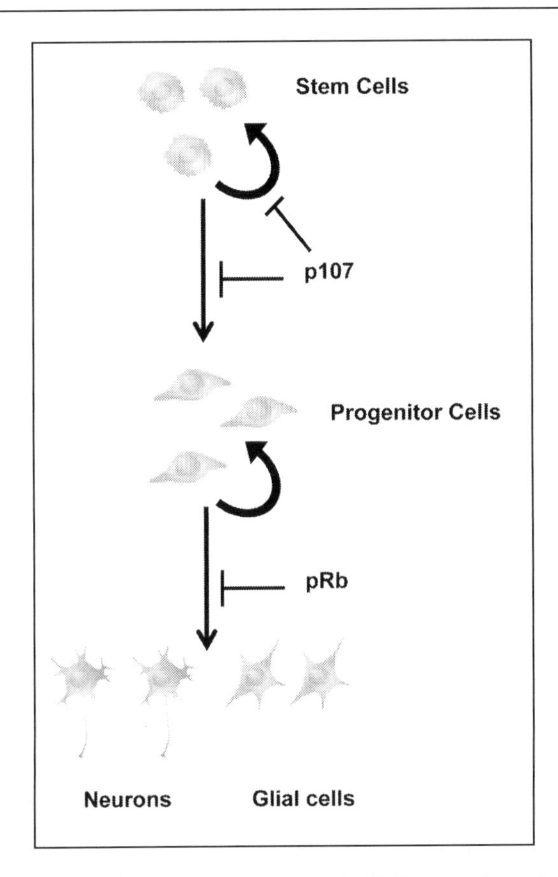

Figure 4. Rb and p107 regulate distinct steps in neurogenesis. P107 negatively regulates proliferation of the neural stem population. In contrast, Rb acts on the neural progenitor pool negatively regulating cell cycle exit and neural differentiation.

in introns 1 and 2 and in the 3' UTR indicated that p107 might directly regulate Notch1 expression at the level of the gene. Chromatin immunoprecipitation and p107 overexpression experiments further supported a role for p107 in directly regulating Notch1 expression.[240] These experiments demonstrate that p107 regulates the neural stem cell pool through the Notch-Hes pathway.

In summary, we have identified a unique role for p107 in the negative regulation of the neural stem cell population. P107 through interaction with Notch signaling regulates stem cell self-renewal thereby affecting the overall numbers of stem and progenitor cells (Fig. 4). This is in sharp contrast to the roles of Rb, which we and others have shown negatively regulates exit of the cell cycle and the onset of terminal differentiation genes.[48,72] Hence, Rb and p107 regulate distinct steps in neurogenesis (Fig. 4).

Rb and Cancer

The Rb gene is inactivated in many human cancers, but it is a component in a pathway that is defective in virtually all neoplasia.[246] A single allele mutation of Rb results in a >90% risk of developing retinoblastoma. Retinoblastoma is an aggressive cancer that left untreated results in loss of vision in newborn children and can be ultimately fatal. Identifying the cell of origin therefore would aid in the treatment and prevention of retinoblastoma.

The study of retinoblastoma has been impeded due to the lack of an animal model. While in humans, loss of *RB* is rate-limiting to the development of retinoblastoma, *RB* heterozygous mice fail to develop retinal tumours, but primarily form thyroid medullary carcinomas and pituitary adenocarcinomas.[110,115,163,254] Rb chimeric mice also fail to develop retinoblastoma, suggesting that Rb loss in the mouse retina is insufficient to induce tumour development.[163,255] The reason why Rb-mediated tumour formation differs between mice and humans is not clear. It has been suggested to be due to species-specific differences in the number of susceptible cells, the timing of susceptibility, or a difference in mutational requirements.[95,102] It may be that additional mutations are required for retinoblastoma development in the mouse. While p107 or p130-deficient mice are not predisposed to tumour formation,[42,146] Rb/p107 double null chimeric mice develop retinoblastoma, suggesting that p107 can act as a tumour suppressor in the absence of Rb,[207] reviewed in ref. 243. In addition, the embryonic lethality of Rb null mice and Rb/p107 double null mice has impeded the generation of a mouse model of inheritable retinoblastoma.

Early mouse models of retinoblastoma relied on the use of viral oncogene inactivation of Rb to overcome the embryonic lethality of germline Rb deletion. Specifically, SV40 large T antigen, human papilloma virus E7, and adenovirus E1A bind and inactivate Rb.[51,62-64,182] Although viral inactivation of Rb successfully resulted in retinal tumours, direct comparisons with retinoblastoma are limited.[5,105] Viral oncogenes are not specific for Rb, but bind and inactivate all three members of the Rb family as well as interacting with other cell cycle proteins such as CKI p21^{Cip1}.[121] In addition, these viral mediated retinal tumours demonstrated the differential sensitivity to loss of Rb in retinal progenitor cells. For instance, E7 mediated Rb loss in photoreceptor cells resulted in apoptosis whereas SV40 large T antigen which inactivates p53 as well as the entire Rb family allowed for oncogenic transformation of photoreceptor cells. These findings demonstrated that photoreceptor cells in mice undergo apoptosis as a result of Rb loss, however in the absence of Rb family members and p53, some photoreceptor cells escaped apoptosis and formed retinal tumours.[106] The enhanced retinal tumour formation by inactivation of the Rb family on a p53 null mouse lead to the “Death Hypothesis”. The hypothesis suggests that retinoblastoma derives from a cell that initially sustains a null mutation in Rb, and in order to survive the loss of Rb it must have an additional mutation leading to protection from apoptosis.[78,106] The death hypothesis however, is less clear in human retinoblastoma since the p53 gene is intact.[78]

The recent development of conditional Rb mouse mutants has enhanced our understanding of the differential cellular sensitivity of Rb loss in a variety of cell types. To generate a conditional Rb knockout, transgenic mice expressing Cre recombinase from a tissue specific promoter were crossed to mice with floxed Rb gene (Rb$^{loxP/loxP}$). Three recent studies have taken advantage of this system to generate mice carrying a conditional knockout of Rb. Using the Pax6 α-enhancer to express Cre recombinase in Pax6-expressing peripheral retinal progenitor cells (Chen and colleagues), were able to selectively knockout Rb expression in the retina at E9.5.[29] The Pax6-Cre; Rb$^{loxp/loxp}$ mice were subsequently interbred with p107 null mice to generate mice carrying mutations for both Rb and p107 in the retina and thereby producing the first mouse model of retinoblastoma. The researchers demonstrated that deletion of Rb and p107 affected cell cycle exit of all cell types in the retina resulting in apoptosis of differentiating rods, cones, ganglion and bipolar cells, whereas the numbers of amacrine, horizontal and Muller glia cells were comparable to that in wild type mice.[29] Significantly, their findings revealed that the cells that survived Rb and p107 loss and ultimately produced retinal tumours, did not require additional mutations to protect them from apoptosis, but were intrinsically death-resistant. Ectopically dividing death-resistant cell types eventually exited the cell cycle as they terminally differentiated and sporadic tumours arose from cells that escaped growth arrest. Retinoblastoma therefore, arose from cells that survived the loss of Rb and p107, underwent extra rounds of cell division due to defective cell cycle exit as a result of Rb loss and acquired a mutation that overcomes growth arrest.[29] These results demonstrated for the first time that the cell-of-origin in retinoblastoma is an intrinsically death-resistant differentiating cell that has an enhanced proliferative capacity.

The second study used the Nestin promoter to express Cre recombinase in the nervous system. Nestin is expressed in all neural progenitors throughout the nervous system; therefore Rb is deleted from all neural progenitor cells. The Nestin-Cre transgene was inserted into an imprinted part of the genome, which resulted in embryonic lethality when Nestin-Cre was inherited paternally, whereas maternal inheritance produced a chimeric/mosaic deletion of Rb and was not lethal.[166] Similar to the Pax6-Cre conditional Rb mutant, the Nestin-Cre Rb mutant revealed that Rb loss during embryonic retinal development, results in ectopic mitosis, and apoptosis of photoreceptors and bipolar cells. Interestingly, when the Nestin-Cre conditional Rb mutants were crossed onto a p53 null background, p53-independent apoptosis was observed in the developing retina and retinoblastoma did not ensue. These results further demonstrated that p53 mutations do not play a role in retinoblastoma indicating a flaw in the 'Death Model'.[166] Retinoblastoma only occurred when the Nestin-Cre conditional mutants were crossed onto a p107 or p130 null background.

A third study used the Chx10 promoter to drive Cre expression and remove Rb from Chx10 expressing cells. Unlike the widespread CNS expression of Nestin, Chx10 is only expressed in the retina and specifically in retinal progenitors, bipolar cells and a subset of Muller glia.[271] The results from this study are consistent with the two previous studies demonstrating that both Rb and p107 must be deleted in the retina for retinoblastoma to occur. All three studies demonstrate Rb deletion alone in the retina is not sufficient and that compensation by p107 must be removed for retinoblastoma to occur in mice.

The results from these studies revealed that the cells responsible for retinoblastoma in mice have two distinguishing characteristics: (i) they have a limited proliferative capacity and (ii) are intrinsically death-resistant (i.e., do not require an additional mutation in the p53 gene). In light of these recent findings from conditional Rb null mice, a new model of retinoblastoma has been proposed: 'the Differentiation Model'.[19,29] Rather than an evasion of death, they proposed that the cell-of-origin in retinoblastoma arises from an intrinsically death-resistant differentiating cell. In the presence of Rb/p107, this cell would normally be post-mitotic but divides ectopically in their absence. This cell is not yet transformed and will eventually undergo growth arrest due to terminal differentiation unless an additional mutation occurs that permits continued proliferation. Note all these studies have been performed using mouse models of retinoblastoma; the next step is to test whether these results are applicable to human retinoblastoma.

Future Directions

Over the past 5 years studies have highlighted new roads to be explored in furthering our understanding of the multiple roles of Rb in cell biology. One path leads to understanding cell specificity, why some cells are nonresponsive to the loss of Rb, whereas others undergo apoptosis and still others fail to exit the cell cycle and continue to proliferate. A second is identifying the role of Rb in terminal mitosis, which genes essential for exiting the cell cycle are disrupted in the absence of Rb. A potential target has recently been identified. Double null mutation of Id2 and Rb rescues the ectopic mitosis phenotype in the Rb deficient mouse. Further research exploring interactions between Rb and the Id pathway need to be elucidated to understand their roles in regulating cell cycle exit. A third points in the direction of identifying the distinct roles of p107 and p130. We have recently demonstrated a unique role for p107 in regulating the neural stem cell population through interactions with the Notch-Hes signaling pathway.[240] It is likely that p130 may function at a much later development time point than either p107 or Rb, consistent with its role in post-mitotic cells.

The generation of mice carrying conditional mutations has enhanced our understanding of the roles of Rb in embryonic development. For instance, conditional deletion of Rb in the telencephalon revealed that the neural apoptotic phenotype observed in the germline Rb mutant was not due to a cell autonomous defect.[72] Most telencephalic neurons could survive in the absence of Rb. Deletion of Rb however did cause a cell autonomous defect in terminal mitosis of neurons resulting in ectopic cell division.[72] In addition, the work by Leone and colleagues demonstrating the importance of Rb expression in extra-embryonic tissue further emphasized the cell intrinsic and noncell intrinsic defects associated with the Rb null phenotype.[48,261] Recently, conditional Rb deletion in the retina has resulted in the identification of the cell-of-origin in retinoblastoma. This finding has changed the direction of research on retinoblastoma from a 'Death Model' where cells were believed to require an additional mutation to become apoptosis resistant to the 'Differentiation Model' where the cell-of-origin is inherently apoptosis resistant and in the absence of Rb undergoes more cell divisions[29] and reviewed in refs. 19, 61. As new techniques emerge and novel interactions are identified with different signaling pathways, the multiple roles of Rb in cell biology will become further elucidated.

Acknowledgements

The authors would like to thank Dr. Rod Bremner for his critical review and helpful input. This work was funded by a Canadian Institutes of Health Research (CIHR) grant to RSS. JLV is a recipient of a Heart & Stroke Foundation of Canada fellowship and KLF is a recipient of a CIHR fellowship and RSS is a CIHR scholar. Jacqueline L. Vanderluit and Kerry L. Ferguson contributed equally.

References

1. Adams MR, Sears R, Nuckolls F et al. Complex transcriptional regulatory mechanisms control expression of the E2F3 locus. Mol Cell Biol 2000; 20:3633-3639.
2. Adams PD. Regulation of the retinoblastoma tumor suppressor protein by cyclin/cdks. Biochim Biophys Acta 2001; 1471:M123-133.
3. Allen KE, de la Luna S, Kerkhoven RM et al. Distinct mechanisms of nuclear accumulation regulate the functional consequence of E2F transcription factors. J Cell Sci 1997; 110(Pt 22):2819-2831.
4. Alt JR, Gladden AB, Diehl JA. p21(Cip1) promotes cyclin D1 nuclear accumulation via direct inhibition of nuclear export. J Biol Chem 2002; 277:8517-8523.
5. al-Ubaidi MR, Font RL, Quiambao AB et al. Bilateral retinal and brain tumors in transgenic mice expressing simian virus 40 large T antigen under control of the human interphotoreceptor retinoid-binding protein promoter. J Cell Biol 1992; 119:1681-1687.
6. Asano M, Nevins JR, Wharton RP. Ectopic E2F expression induces S phase and apoptosis in Drosophila imaginal discs. Genes Dev 1996; 10:1422-1432.
7. Bagchi S, Weinmann R, Raychaudhuri P. The retinoblastoma protein copurifies with E2F-I, an E1A-regulated inhibitor of the transcription factor E2F. Cell 1991; 65:1063-1072.
8. Bandara LR, La Thangue NB. Adenovirus E1A prevents the retinoblastoma gene product from complexing with a cellular transcription factor. Nature 1991; 351:494-497.
9. Beijersbergen RL, Carlee L, Kerkhoven RM et al. Regulation of the retinoblastoma protein-related p107 by G1 cyclin complexes. Genes Dev 1995; 9:1340-1353.
10. Beijersbergen RL, Kerkhoven RM, Zhu L et al. E2F-4, a new member of the E2F gene family, has oncogenic activity and associates with p107 in vivo. Genes Dev 1994; 8:2680-2690.
11. Benezra R, Davis RL, Lockshon D et al. The protein Id: A negative regulator of helix-loop-helix DNA binding proteins. Cell 1990; 61:49-59.
12. Bernards R, Schackleford GM, Gerber MR et al. Structure and expression of the murine retinoblastoma gene and characterization of its encoded protein. Proc Natl Acad Sci USA 1989; 86:6474-6478.
13. Blain SW, Montalvo E, Massague J. Differential interaction of the cyclindependent kinase (Cdk) inhibitor p27Kip1 with cyclin A-Cdk2 and cyclin D2Cdk4. J Biol Chem 1997; 272:25863-25872.
14. Bookstein R, Rio P, Madreperla SA et al. Promoter deletion and loss of retinoblastoma gene expression in human prostate carcinoma. Proc Natl Acad Sci USA 1990; 87:7762-7766.
15. Bookstein R, Lee EY, To H et al. Human retinoblastoma susceptibility gene: Genomic organization and analysis of heterozygous intragenic deletion mutants. Proc Natl Acad Sci USA 1988; 85:2210-2214.
16. Botz J, Zerfass-Thome K, Spitkovsky D et al. Cell cycle regulation of the murine cyclin E gene depends on an E2F binding site in the promoter. Mol Cell Biol 1996; 16:3401-3409.
17. Bracken AP, Ciro M, Cocito A et al. E2F target genes: Unraveling the biology. Trends Biochem Sci 2004; 29:409-417.

18. Brehm A, Miska EA, McCance DJ et al. Retinoblastoma protein recruits histone deacetylase to repress transcription. Nature 1998; 391:535-536.
19. Bremner R, Chen D, Pacal M et al. The Rb protein family in retinal development and retinoblastoma: New insights from mouse models. Developmental Neuroscience 2004; in press.
20. Bremner R, Cohen BL, Sopta M et al. Direct transcriptional repression by pRB and its reversal by specific cyclins. Mol Cell Biol 1995; 15:3256-3265.
21. Bruce JL, Hurford Jr RK, Classon M et al. Requirements for cell cycle arrest by p16INK4a. Mol Cell 2000; 6:737-742.
22. Buchkovich K, Duffy LA, Harlow E. The retinoblastoma protein is phosphorylated during specific phases of the cell cycle. Cell 1989; 58:1097-1105.
23. Buck V, Allen KE, Sorensen T et al. Molecular and functional characterisation of E2F-5, a new member of the E2F family. Oncogene 1995; 11:31-38.
24. Busser J, Geldmacher DS, Herrup K. Ectopic cell cycle proteins predict the sites of neuronal cell death in Alzheimer's disease brain. J Neurosci 1998; 18:2801-2807.
25. Callaghan DA, Dong L, Callaghan SM et al. Neural precursor cells differentiating in the absence of Rb exhibit delayed terminal mitosis and deregulated E2F 1 and 3 activity. Dev Biol 1999; 207:257-270.
26. Castano E, Kleyner Y, Dynlacht BD. Dual cyclin-binding domains are required for p107 to function as a kinase inhibitor. Mol Cell Biol 1998; 18:5380-5391.
27. Chao R, Khan W, Hannun YA. Retinoblastoma protein dephosphorylation induced by D-erythro-sphingosine. J Biol Chem 1992; 267:23459-23462.
28. Chellappan S, Hiebert S, Mudryj M et al. The E2F transcription factor is a cellular target for the Rb protein. Cell 1991; 65:1053-1061.
29. Chen D, Livne-bar I, Vanderluit JL et al. Cell-specific effects of RB or RB/p107 loss on retinal development implicate an intrinsically death-resistant cell-of-origin in retinoblastoma. Cancer Cell 2004; 5:539-551.
30. Chen PL, Riley DJ, Chen Y et al. Retinoblastoma protein positively regulates terminal adipocyte differentiation through direct interaction with C/EBPs. Genes Dev 1996; 10:2794-2804.
31. Chen PL, Scully P, Shew JY et al. Phosphorylation of the retinoblastoma gene product is modulated during the cell cycle and cellular differentiation. Cell 1989; 58:1193-1198.
32. Chen TT, Wang JY. Establishment of irreversible growth arrest in myogenic differentiation requires the RB LXCXE-binding function. Mol Cell Biol 2000; 20:5571-5580.
33. Cheng M, Olivier P, Diehl JA et al. The p21(Cip1) and p27(Kip1) CDK 'inhibitors' are essential activators of cyclin D-dependent kinases in murine fibroblasts. Embo J 1999; 18:1571-1583.
34. Chittenden T, Livingston DM, Kaelin Jr WG. The T/E1A-binding domain of the retinoblastoma product can interact selectively with a sequence-specific DNA-binding protein. Cell 1991; 65:1073-1082.
35. Chow KN, Dean DC. Domains A and B in the Rb pocket interact to form a transcriptional repressor motif. Mol Cell Biol 1996; 16:4862-4868.
36. Clarke AR, Maandag ER, van Roon M et al. Requirement for a functional Rb-1 gene in murine development. Nature 1992; 359:328-330.
37. Classon M, Dyson N. p107 and p130: Versatile proteins with interesting pockets. Exp Cell Res 2001; 264:135-147.
38. Classon M, Kennedy BK, Mulloy R et al. Opposing roles of pRB and p107 in adipocyte differentiation. Proc Natl Acad Sci USA 2000a; 97:10826-10831.
39. Classon M, Salama S, Gorka C et al. Combinatorial roles for pRB, p107, and p130 in E2F-mediated cell cycle control. Proc Natl Acad Sci USA 2000b; 97:10820-10825.
40. Claudio PP, Howard CM, Baldi A et al. p130/pRb2 has growth suppressive properties similar to yet distinctive from those of retinoblastoma family members pRb and p107. Cancer Res 1994; 54:5556-5560.
41. Cobrinik D, Whyte P, Peeper DS et al. Cell cycle-specific association of E2F with the p130 E1A-binding protein. Genes Dev 1993; 7:2392-2404.
42. Cobrinik D, Lee MH, Hannon G et al. Shared role of the pRB-related p130 and p107 proteins in limb development. Genes Dev 1996; 10:1633-1644.
43. Coqueret O. Linking cyclins to transcriptional control. Gene 2002; 299:35-55.
44. Dahiya A, Gavin MR, Luo RX et al. Role of the LXCXE binding site in Rb function. Mol Cell Biol 2000; 20:6799-6805.
45. Dalton S. Cell Cycle regulation of the human cdc 2 gene. EMBO Journal 1992; 11:1797-1807.
46. Dannenberg JH, van Rossum A, Schuijff L et al. Ablation of the retinoblastoma gene family deregulates G(1) control causing immortalization and increased cell turnover under growth-restricting conditions. Genes Dev 2000; 14:3051-3064.

47. de Bruin A, Maiti B, Jakoi L et al. Identification and characterization of E2F7, a novel mammalian E2F family member capable of blocking cellular proliferation. J Biol Chem 2003; 278:42041-42049.
48. De Bruin A, Wu L, Saavedra HI et al. Rb function in extraembryonic lineages suppresses apoptosis in the CNS of Rb-deficient mice. Proc Natl Acad Sci USA 2003; 100:6546-6551.
49. de la Luna S, Burden MJ, Lee CW et al. Nuclear accumulation of the E2F heterodimer regulated by subunit composition and alternative splicing of a nuclear localization signal. J Cell Sci 1996; 109(Pt 10):2443-2452.
50. DeCaprio JA, Ludlow JW, Lynch D et al. The product of the retinoblastoma susceptibility gene has properties of a cell cycle regulatory element. Cell 1989; 58:1085-1095.
51. DeCaprio JA, Ludlow JW, Figge J et al. SV40 large tumor antigen forms a specific complex with the product of the retinoblastoma susceptibility gene. Cell 1988; 54:275-283.
52. deGregori J, Kowalik T, Nevins JR. Cellular targets for activation by the E2F1 transcription factor include DNA synthesis and G1/S -regulatory genes. Mol Cell Biol 1995; 15:4215-4224.
53. DeGregori J, Leone G, Ohtani K et al. E2F-1 accumulation bypasses a G1 arrest resulting from the inhibition of G1 cyclin-dependent kinase activity. Genes Dev 1995; 9:2873-2887.
54. deGregori J, Leone G, Miron A et al. Distinct roles for E2F proteins in cell growth control and apoptosis. Proc Natl Acad Sci USA 1997; 94:7245-7250.
55. Di Stefano L, Jensen MR, Helin K. E2F7, a novel E2F featuring DP-independent repression of a subset of E2F-regulated genes. Embo J 2003; 22:6289-6298.
56. Dick FA, Dyson N. pRB contains an E2F1-specific binding domain that allows E2F1-induced apoptosis to be regulated separately from other E2F activities. Mol Cell 2003; 12:639-649.
57. D'Souza SJ, Pajak A, Balazsi K et al. Ca2+ and BMP-6 signaling regulate E2F during epidermal keratinocyte differentiation. J Biol Chem 2001; 276:23531-23538.
58. Dulic V, Lees E, Reed SI. Association of human cyclin E with a periodic G1-S phase protein kinase. Science 1992; 257:1958-1961.
59. Dunaief JL, Strober BE, Guha S et al. The retinoblastoma protein and BRG1 form a complex and cooperate to induce cell cycle arrest. Cell 1994; 79:119-130.
60. Dunn JM, Phillips RA, Becker AJ et al. Identification of germline and somatic mutations affecting the retinoblastoma gene. Science 1988; 241:1797-1800.
61. Dyer MA, Bremner R. Cell-of-origin in cancer initiation and progression: Lessons from Retinoblastoma. Nat Rev Cancer 2004; in press.
62. Dyson N. The regulation of E2F by pRb-family proteins. Genes Dev 1998; 12:2245-2262.
63. Dyson N, Howley PM, Munger K et al. The human papilloma virus-16 E7 oncoprotein is able to bind to the retinoblastoma gene product. Science 1989; 243:934-937.
64. Dyson N, Bernards R, Friend SH et al. Large T antigens of many polyomaviruses are able to form complexes with the retinoblastoma protein. J Virol 1990; 64:1353-1356.
65. Egan C, Bayley ST, Branton PE. Binding of the Rb1 protein to E1A products is required for adenovirus transformation. Oncogene 1989; 4:383-388.
66. Eng C, Li FP, Abramson DH et al. Mortality from second tumors among long-term survivors of retinoblastoma. J Natl Cancer Inst 1993; 85:1121-1128.
67. Estivill-Torrus G, Pearson H, van Heyningen V et al. Pax6 is required to regulate the cell cycle and the rate of progression from symmetrical to asymmetrical division in mammalian cortical progenitors. Development 2002; 129:455-466.
68. Ewen ME, Xing YG, Lawrence JB et al. Molecular cloning, chromosomal mapping, and expression of the cDNA for p107, a retinoblastoma gene product-related protein. Cell 1991; 66:1155-1164.
69. Ewen ME, Faha B, Harlow E et al. Interaction of p107 with cyclin A independent of complex formation with viral oncoproteins. Science 1992; 255:85-87.
70. Faha B, Ewen ME, Tsai LH et al. Interaction between human cyclin A and adenovirus E1A-associated p107 protein. Science 1992; 255:87-90.
71. Fajas L, Landsberg RL, Huss-Garcia Y et al. E2Fs regulate adipocyte differentiation. Dev Cell 2002; 3:39-49.
72. Ferguson KL, Vanderluit JL, Hebert JM et al. Telencephalon-specific Rb knockouts reveal enhanced neurogenesis, survival and abnormal cortical development. Embo J 2002; 21:3337-3346.
73. Flemington EK, Speck SH, Kaelin Jr WG. E2F-1-mediated transactivation is inhibited by complex formation with the retinoblastoma susceptibility gene product. Proc Natl Acad Sci USA 1993; 90:6914-6918.
74. Fortin A, MacLaurin JG, Arbour N et al. The proapoptotic gene SIVA is a direct transcriptional target for the tumor suppressors p53 and E2F1. J Biol Chem 2004; 279:28706-28714.
75. Freeman RS, Estus S, Johnson Jr EM. Analysis of cell cycle-related gene expression in postmitotic neurons: Selective induction of Cyclin D1 during programmed cell death. Neuron 1994; 12:343-355.

76. Friend SH, Bernards R, Rogelj S et al. A human DNA segment with properties of the gene that predisposes to retinoblastoma and osteosarcoma. Nature 1986; 323:643-646.
77. Fung YK, Murphree AL, T'Ang A et al. Structural evidence for the authenticity of the human retinoblastoma gene. Science 1987; 236:1657-1661.
78. Gallie BL, Campbell C, Devlin H et al. Developmental basis of retinal-specific induction of cancer by RB mutation. Cancer Res 1999; 59:1731s-1735s.
79. Gallie BL, Squire JA, Goddard A et al. Mechanism of oncogenesis in retinoblastoma. Lab Invest 1990; 62:394-408.
80. Gao CY, Zelenka PS. Induction of cyclin B and H1 kinase activity in apoptotic PC12 cells. Exp Cell Res 1995; 219:612-618.
81. Gaubatz S, Lees JA, Lindeman GJ et al. E2F4 is exported from the nucleus in a CRM1-dependent manner. Mol Cell Biol 2001; 21:1384-1392.
82. Ginsberg D, Vairo G, Chittenden T et al. E2F-4, a new member of the E2F transcription factor family, interacts with p107. Genes Dev 1994; 8:2665-2679.
83. Giovanni A, Keramaris E, Morris EJ et al. E2F1 mediates death of B-amyloid-treated cortical neurons in a manner independent of p53 and dependent on Bax and caspase 3. J Biol Chem 2000; 275:11553-11560.
84. Girard F, Strausfeld U, Fernandez A et al. Cyclin A is required for the onset of DNA replication in mammalian fibroblasts. Cell 1991; 67:1169-1179.
85. Goodrich DW, Wang NP, Qian YW et al. The retinoblastoma gene product regulates progression through the G1 phase of the cell cycle. Cell 1991; 67:293-302.
86. Guo Z, Yikang S, Yoshida H et al. Inactivation of the retinoblastoma tumor suppressor induces apoptosis protease-activating factor-1 dependent and independent apoptotic pathways during embryogenesis. Cancer Res 2001; 61:8395-8400.
87. Hamel PA, Gill RM, Phillips RA et al. Transcriptional repression of the E2containing promoters EIIaE, c-myc, and RB1 by the product of the RB1 gene. Mol Cell Biol 1992; 12:3431-3438.
88. Hannon GJ, Demetrick D, Beach D. Isolation of the Rb-related p130 through its interaction with CDK2 and cyclins. Genes Dev 1993; 7:2378-2391.
89. Harbour JW. Overview of RB gene mutations in patients with retinoblastoma. Implications for clinical genetic screening. Ophthalmology 1998; 105:1442-1447.
90. Harbour JW, Dean DC. The Rb/E2F pathway: Expanding roles and emerging paradigms. Genes Dev 2000; 14:2393-2409.
91. Harbour JW, Lai SL, Whang-Peng J et al. Abnormalities in structure and expression of the human retinoblastoma gene in SCLC. Science 1988; 241:353-357.
92. Helin K, Harlow E. The retinoblastoma protein as a transcriptional repressor. Trends Cell Biol 1993; 3:43-45.
93. Helin K, Harlow E, Fattaey A. Inhibition of E2F-1 transactivation by direct binding of the retinoblastoma protein. Mol Cell Biol 1993; 13:6501-6508.
94. Helin K, Lees JA, Vidal M et al. A cDNA encoding a pRB-binding protein with properties of the transcription factor E2F. Cell 1992; 70:337-350.
95. Herrera RE, Sah VP, Williams BO et al. Altered cell cycle kinetics, gene expression, and G1 restriction point regulation in Rbdeficient fibroblasts. Mol Cell Biol 1996; 16:2402-2407.
96. Hershko T, Ginsberg D. Up-regulation of BH3-only proteins by E2F1: Role in apoptosis. J Biol Chem 2003.
97. Hiebert SW, Lipp M, Nevins JR. E1A-dependent trans-activation of the human MYC promoter is mediated by the E2F factor. Proc Natl Acad Sci USA 1989; 86:3594-3598.
98. Hijmans EM, Voorhoeve PM, Beijersbergen RL et al. E2F5, a new E2F family member that interacts with p130 in vivo. Mol Cell Biol 1995; 15:3082-3089.
99. Hinds PW, Mittnacht S, Dulic V et al. Regulation of retinoblastoma protein functions by ectopic expression of human cyclins. Cell 1992; 70:993-1006.
100. Hitoshi S, Alexson T, Tropepe V et al. Notch pathway molecules are essential for the maintenance, but not the generation, of mammalian neural stem cells. Genes Dev 2002; 16:846-858.
101. Hong FD, Huang HJ, To H et al. Structure of the human retinoblastoma gene. Proc Natl Acad Sci USA 1989; 86:5502-5506.
102. Hooper ML. Tumour suppressor gene mutations in humans and mice: Parallels and contrasts. Embo J 1998; 17:6783-6789.
103. Horowitz JM, Park SH, Bogenmann E et al. Frequent inactivation of the retinoblastoma anti-oncogene is restricted to a subset of human tumor cells. Proc Natl Acad Sci USA 1990; 87:2775-2779.
104. Hou ST, Callaghan D, Fournier MC et al. The transcription factor E2F1 modulates apoptosis of neurons. J Neurochem 2000; 75:91-100.

105. Howes KA, Lasudry JG, Albert DM et al. Photoreceptor cell tumors in transgenic mice. Invest Ophthalmol Vis Sci 1994a; 35:342-351.
106. Howes KA, Ransom N, Papermaster DS et al. Apoptosis or retinoblastoma: Alternative fates of photoreceptors expressing the HPV-16 E7 gene in the presence or absence of p53. Genes Dev 1994b; 8:1300-1310.
107. Hsiao KM, McMahon SL, Farnham PJ. Multiple DNA elements are required for the growth regulation of the mouse E2Fl promoter. Genes Dev 1994; 8:1526-1537.
108. Hsieh JK, Fredersdorf S, Kouzarides T et al. E2F1-induced apoptosis requires DNA binding but not transactivation and is inhibited by the retinoblastoma protein through direct interaction. Genes Dev 1997; 11:1840-1852.
109. Hu N, Gulley ML, Kung JT et al. Retinoblastoma gene deficiency has mitogenic but not tumorigenic effects on erythropoiesis. Cancer Res 1997; 57:4123-4129.
110. Hu N, Gutsmann A, Herbert DC et al. Heterozygous Rb1 delta 20/+mice are predisposed to tumors of the pituitary gland with a nearly complete penetrance. Oncogene 1994; 9:1021-1027.
111. Huang HJ, Yee JK, Shew JY et al. Suppression of the neoplastic phenotype by replacement of the RB gene in human cancer cells. Science 1988; 242:1563-1566.
112. Hurford RK, Cobrinik D, Lee MH et al. pRb and p107/p130 are required for the regulated expression of different sets of E2F responsive genes. Genes Dev 1997; 11:1447-1463.
113. Iavarone A, Garg P, Lasorella A et al. The helix-loop-helix protein Id-2 enhances cell proliferation and binds to the retinoblastoma protein. Genes Dev 1994; 8:1270-1284.
114. Jacks T. Tumor suppressor gene mutations in mice. Annu Rev Genet 1996; 30:603-636.
115. Jacks T, Fazeli A, Schmitt EM et al. Effects of an Rb mutation in the mouse. Nature 1992; 359:295-300.
116. Jan YN, Jan LY. HLH proteins, fly neurogenesis, and vertebrate myogenesis. Cell 1993; 75:827-830.
117. Jiang Z, Zacksenhaus E, Gallie BL et al. The retinoblastoma gene family is differentially expressed during embryogenesis. Oncogene 1997; 14:1789-1797.
118. Jiang Z, Liang P, Leng R et al. E2F1 and p53 are dispensable, whereas p21(Waf1/Cip1) cooperates with Rb to restrict endoreduplication and apoptosis during skeletal myogenesis. Dev Biol 2000; 227:8-41.
119. Johnson DG, Ohtani K, Nevins JR. Autoregulatory control of E2F1 expression in response to positive and negative regulators of cell cycle progression. Genes Dev 1994; 8:1514-1525.
120. Johnson DG, Schwarz JK, Cress WD et al. Expression of transcription factor E2F1 induces quiescent cells to enter S phase. Nature 1993; 365:349-352.
121. Jones DL, Alani RM, Munger K. The human papillomavirus E7 oncoprotein can uncouple cellular differentiation and proliferation in human keratinocytes by abrogating p21Cip1-mediated inhibition of cdk2. Genes Dev 1997; 11:2101-2111.
122. Kaelin Jr WG, Krek W, Sellers WR et al. Expression cloning of a cDNA encoding a retinoblastoma-binding protein with E2F-like properties. Cell 1992; 70:351-364.
123. Karlseder J, Rotheneder H, Wintersberger E. Interaction of Sp1 with the growth- and cell cycle-regulated transcription factor E2F. Mol Cell Biol 1996; 16:1659-1667.
124. Kiess M, Gill RM, Hamel PA. Expression and activity of the retinoblastoma protein (pRB)-family proteins, p107 and p130, during L6 myoblast differentiation. Cell Growth Differ 1995; 6:1287-1298.
125. Kim HY, Cho Y. Structural similarity between the pocket region of retinoblastoma tumour suppressor and the cyclin-box. Nat Struct Biol 1997; 4:390-395.
126. Kingston RE, Narlikar GJ. ATP-dependent remodeling and acetylation as regulators of chromatin fluidity. Genes Dev 1999; 13:2339-2352.
127. Knudsen ES, Wang JY. Differential regulation of retinoblastoma protein function by specific Cdk phosphorylation sites. J Biol Chem 1996; 271:8313-8320.
128. Knudson Jr AG. Mutation and cancer: Statistical study of retinoblastoma. Proc Natl Acad Sci USA 1971; 68:820-823.
129. Koff A, Giordano A, Desai D et al. Formation and activation of a cyclin E-CDK2 complex during the G1 phase of the human cell cycle. Science 1992; 257:1689-1694.
130. Kornberg RD, Lorch Y. Twenty-five years of the nucleosome, fundamental particle of the eukaryote chromosome. Cell 1999; 98:285-294.
131. Kowalik TF, DeGregori J, Schwarz JK et al. E2F1 overexpression in quiescent fibroblasts leads to induction of cellular DNA synthesis and apoptosis. J Virol 1995; 69:2491-2500.
132. LaBaer J, Garrett MD, Stevenson LF et al. New functional activities for the p21 family of CDK inhibitors. Genes Dev 1997; 11:847-862.
133. Lai A, Marcellus RC, Corbeil HB et al. RBP1 induces growth arrest by repression of E2F-dependent transcription. Oncogene 1999a; 18:2091-2100.

134. Lai A, Lee JM, Yang WM et al. RBP1 recruits both histone deacetylase-dependent and -independent repression activities to retinoblastoma family proteins. Mol Cell Biol 1999b; 19:6632-6641.
135. Lam EW, La Thangue NB. DP and E2F proteins: Coordinating transcription with cell cycle progression. Curr Opin Cell Biol 1994; 6:859-866.
136. Lam EWF, Watson RJ. An E2F-binding site mediates cell cycle regulated repression of mouse B-myb transcription. EMBO Journal 1993; 12:2705-2713.
137. Landsberg RL, Sero JE, Danielian PS et al. The role of E2F4 in adipogenesis is independent of its cell cycle regulatory activity. Proc Natl Acad Sci USA 2003; 100:2456-2461.
138. Lasorella A, Iavarone A, Israel MA. Id2 specifically alters regulation of the cell cycle by tumor suppressor proteins. Mol Cell Biol 1996; 16:2570-2578.
139. Lasorella A, Noseda M, Beyna M et al. Id2 is a retinoblastoma protein target and mediates signalling by Myc oncoproteins. Nature 2000; 407:592-598.
140. LeCouter JE, Kablar B, Whyte PF et al. Strain-dependent embryonic lethality in mice lacking the retinoblastoma-related p130 gene. Development 1998a; 125:4669-4679.
141. LeCouter JE, Kablar B, Hardy WR et al. Strain-dependent myeloid hyperplasia, growth deficiency, and accelerated cell cycle in mice lacking the Rb-related p107 gene. Mol Cell Biol 1998b; 18:7455-7465.
142. Lee E, Hu N, Yuan SSF et al. Dual roles of the retinoblastoma protein in cell cycle regulation and neuron differentiation. Genes & Dev 1994; 8:2008-2021.
143. Lee EY, To H, Shew JY et al. Inactivation of the retinoblastoma susceptibility gene in human breast cancers. Science 1988; 241:218-221.
144. Lee EY, Chang CY, Hu N et al. Mice deficient for Rb are nonviable and show defects in neurogenesis and haematopoiesis. Nature 1992; 359:288-294.
145. Lee JO, Russo AA, Pavletich NP. Structure of the retinoblastoma tumour-suppressor pocket domain bound to a peptide from HPV E7. Nature 1998; 391:859-865.
146. Lee MH, Williams BO, Mulligan G et al. Targeted disruption of p107: Functional overlap between p107 and Rb. Genes Dev 1996; 10:1621-1632.
147. Lee WH, Bookstein R, Hong F et al. Human retinoblastoma susceptibility gene: Cloning, identification, and sequence. Science 1987a; 235:1394-1399.
148. Lee WH, Shew JY, Hong FD et al. The retinoblastoma susceptibility gene encodes a nuclear phosphoprotein associated with DNA binding activity. Nature 1987b; 329:642-645.
149. Lees E, Faha B, Dulic V et al. Cyclin E/cdk2 and cyclin A/cdk2 kinases associate with p107 and E2F in a temporally distinct manner. Genes Dev 1992; 6:1874-1885.
150. Lees JA, Saito M, Vidal M et al. The retinoblastoma protein binds to a family of E2F transcription factors. Mol Cell Biol 1993; 13:7813-7825.
151. Leone G, DeGregori J, Yan Z et al. E2F3 activity is regulated during the cell cycle and is required for the induction of S phase. Genes Dev 1998; 12:2120-2130.
152. Leone G, Nuckolls F, Ishida S et al. Identification of a novel E2F3 product suggests a mechanism for determining specificity of repression by Rb proteins. Mol Cell Biol 2000; 20:3626-3632.
153. Li Y, Chopp M, Powers C et al. Immunoreactivity of cyclin D1/cdk4 in neurons and oligodendrocytes after focal cerebral ischemia in rat. J Cereb Blood Flow Metab 1997; 17:846-856.
154. Li Y, Graham C, Lacy S et al. The adenovirus E1A-associated 130-kD protein is encoded by a member of the retinoblastoma gene family and physically interacts with cyclins A and E. Genes Dev 1993; 7:2366-2377.
155. Lindeman GJ, Gaubatz S, Livingston DM et al. The subcellular localization of E2F-4 is cell-cycle dependent. Proc Natl Acad Sci USA 1997; 94:5095-5100.
156. Lindeman GJ, Dagnino L, Gaubatz S et al. A specific, nonproliferative role for E2F-5 in choroid plexus function revealed by gene targeting. Genes Dev 1998; 12:1092-1098.
157. Lipinski MM, Jacks T. The retinoblastoma gene family in differentiation and development. Oncogene 1999; 18:7873-7882.
158. Lipinski MM, Macleod KF, Williams BO et al. Cell-autonomous and noncell-autonomous functions of the Rb tumor suppressor in developing central nervous system. Embo J 2001; 20:3402-3413.
159. Ludlow JW, Glendening CL, Livingston DM et al. Specific enzymatic dephosphorylation of the retinoblastoma protein. Mol Cell Biol 1993; 13:367-372.
160. Ludlow JW, DeCaprio JA, Huang CM et al. SV40 large T antigen binds preferentially to an underphosphorylated member of the retinoblastoma susceptibility gene product family. Cell 1989; 56:57-65.
161. Lukas J, Petersen BO, Holm K et al. Deregulated expression of E2F family members induces S-phase entry and overcomes p16INK4A-mediated growth suppression. Mol Cell Biol 1996; 16:1047-1057.

162. Lundberg AS, Weinberg RA. Functional inactivation of the retinoblastoma protein requires sequential modification by at least two distinct cyclin-cdk complexes. Mol Cell Biol 1998; 18:753-761.
163. Maandag EC, van der Valk M, Vlaar M et al. Developmental rescue of an embryonic-lethal mutation in the retinoblastoma gene in chimeric mice. Embo J 1994; 13:4260-4268.
163a. Luo RX, Postigo AA, Dean DC. Rb interacts with histone deacetylase to repress transcription. Cell 1998l 92:463-473.
164. Macleod KF, Hu Y, Jacks T. Loss of Rb activates both p53-dependent and independent cell death pathways in the developing mouse nervous system. Embo J 1996; 15:6178-6188.
165. MacPherson D, Sage J, Crowley D et al. Conditional mutation of Rb causes cell cycle defects without apoptosis in the central nervous system. Mol Cell Biol 2003; 23:1044-1053.
166. MacPherson D, Sage J, Kim T et al. Cell type-specific effects of Rb deletion in the murine retina. Genes Dev 2004; 18:1681-1694.
167. Magae J, Wu CL, Illenye S et al. Nuclear localization of DP and E2F transcription factors by heterodimeric partners and retinoblastoma protein family members. J Cell Sci 1996; 109(Pt 7):1717-1726.
168. Magnaghi-Jaulin L, Groisman R, Naguibneva I et al. Retinoblastoma protein represses transcription by recruiting a histone deacetylase. Nature 1998; 391:533.
169. Mann DJ, Jones NC. E2F-1 but not E2F-4 can overcome p16-induced G1 cell-cycle arrest. Curr Biol 1996; 6:474-483.
170. Marino S, Hoogervoorst D, Brandner S et al. Rb and p107 are required for normal cerebellar development and granule cell survival but not for Purkinje cell persistence. Development 2003; 130:3359-3368.
171. Mayol X, Grana X, Baldi A et al. Cloning of a new member of the retinoblastoma gene family (pRb2) which binds to the E1A transforming domain. Oncogene 1993; 8:2561-2566.
172. McShea A, Harris PL, Webster KR et al. Abnormal expression of the cell cycle regulators P16 and CDK4 in Alzheimer's disease. Am J Pathol 1997; 150:1933-1939.
173. Mihara K, Cao XR, Yen A et al. Cell cycle-dependent regulation of phosphorylation of the human retinoblastoma gene product. Science 1989; 246:1300-1303.
174. Moberg K, Starz MA, Lees JA. E2F-4 switches from p130 to p107 and pRB in response to cell cycle reentry. Mol Cell Biol 1996; 16:1436-1449.
175. Moll AC, Imhof SM, Bouter LM et al. Second primary tumors in patients with retinoblastoma. A review of the literature. Ophthalmic Genet 1997; 18:27-34.
176. Morgenbesser SD, Williams BO, Jacks T et al. p53-dependent apoptosis produced by Rb-deficiency in the developing mouse lens. Nature 1994; 371:72-74.
177. Morkel M, Wenkel J, Bannister AJ et al. An E2F-like repressor of transcription. Nature 1997; 390:567-568.
178. Moroni MC, Hickman ES, Denchi EL et al. Apaf-1 is a transcriptional target for E2F and p53. Nat Cell Biol 2001; 3:552-558.
179. Morris EJ, Dyson NJ. Retinoblastoma protein partners. Adv Cancer Res 2001; 82:1-54.
180. Mudryj M, Hiebert SW, Nevins JR. A role for the adenovirus inducible E2F transcription factor in a proliferation dependent signal transduction pathway. Embo J 1990; 9:2179-2184.
181. Mulligan G, Jacks T. The retinoblastoma gene family: Cousins with overlapping interests. Trends Genet 1998; 14:223-229.
182. Munger K, Werness BA, Dyson N et al. Complex formation of human papillomavirus E7 proteins with the retinoblastoma tumor suppressor gene product. Embo J 1989; 8:4099-4105.
183. Muraoka RS, Lenferink AE, Law B et al. ErbB2/Neu-induced, cyclin D1-dependent transformation is accelerated in p27-haploinsufficient mammary epithelial cells but impaired in p27-null cells. Mol Cell Biol 2002; 22:2204-2219.
184. Nakamura Y, Sakakibara S, Miyata T et al. The bHLH gene hes1 as a repressor of the neuronal commitment of CNS stem cells. J Neurosci 2000; 20:283-293.
185. Neuman E, Flemington EK, Sellers WR et al. Transcription of the E2F1 gene is rendered cell cycle dependent by E2F DNA-binding sites within its promoter. Mol Cell Biol 1994; 14:6607-6615.
186. Neuman E, Flemington EK, Sellers WR et al. Transcription of the E2F1 gene is rendered cell cycle dependent by E2F DNA-binding sites within its promoter. Mol Cell Biol 1995; 15:4660.
187. Nevins JR. E2F: A link between the retinoblastoma tumor suppressor protein and viral oncoproteins. Science 1992; 258:424-429.
188. Norton JD. ID helix-loop-helix proteins in cell growth, differentiation and tumorigenesis. J Cell Sci 2000; 113(Pt 22):3897-3905.
189. Novitch BG, Mulligan GJ, Jacks T et al. Skeletal muscle cells lacking the retinoblastoma protein display defects in muscle gene expression and accumulate in S and G2 phases of the cell cycle. J Cell Biol 1996; 135:441-456.

190. O'Hare MJ, Hou ST, Morris EJ et al. Induction and modulation of cerebellar granule neuron death by E2F-1. J Biol Chem 2000; 275:25358-25364.
191. Ohtsubo M, Roberts JM. Cyclin-dependent regulation of G1 in mammalian fibroblasts. Science 1993; 259:1908-1912.
192. Ohtsuka T, Sakamoto M, Guillemot F et al. Roles of the basic helix-loop-helix genes Hes1 and Hes5 in expansion of neural stem cells of the developing brain. J Biol Chem 2001; 276:30467-30474.
193. Osuga H, Osuga S, Wang F et al. Cyclin-dependent kinases as a therapeutic target for stroke. Proc Natl Acad Sci USA 2000; 97:10254-10259.
194. Pagano M, Pepperkok R, Verde F et al. Cyclin A is required at two points in the human cell cycle. Embo J 1992; 11:961-971.
195. Pan H, Yin C, Dyson NJ et al. Key roles for E2F1 in signaling p53-dependent apoptosis and in cell division within developing tumors. Mol Cell 1998; 2:283-292.
196. Paramio JM, Segrelles C, Casanova ML et al. Opposite functions for E2F1 and E2F4 in human epidermal keratinocyte differentiation. J Biol Chem 2000; 275:41219-41226.
197. Park DS, Morris EJ, Padmanabhan J et al. Cyclin-dependent kinases participate in death of neurons evoked by DNA-damaging agents. J Cell Biol 1998; 143:457-467.
198. Park DS, Morris EJ, Bremner R et al. Involvement of retinoblastoma family members and E2F/DP complexes in the death of neurons evoked by DNA damage. J Neurosci 2000; 20:3104-3114.
199. Persengiev SP, Kondova II, Kilpatrick DL. E2F4 actively promotes the initiation and maintenance of nerve growth factor-induced cell differentiation. Mol Cell Biol 1999; 19:6048-6056.
200. Phillips AC, Bates S, Ryan KM et al. Induction of DNA synthesis and apoptosis are separable functions of E2F-1. Genes Dev 1997; 11:1835-1863.
201. Porcher C, Swat W, Rockwell K et al. The T cell leukemia oncoprotein SCL/tal-1 is essential for development of all hematopoietic lineages. Cell 1996; 86:47-57.
202. Qin XQ, Chittenden T, Livingston DM et al. Identification of a growth suppression domain within the retinoblastoma gene product. Genes Dev 1992; 6:953-964.
203. Qin XQ, Livingston DM, Kaelin Jr WG et al. Deregulated transcription factor E2F-1 expression leads to S-phase entry and p53-mediated apoptosis. Proc Natl Acad Sci USA 1994; 91:10918-10922.
204. Qin XQ, Livingston DM, Ewen M et al. The transcription factor E2F-1 is a downstream target of RB action. Mol Cell Biol 1995; 15:742-755.
205. Rayman JB, Takahashi Y, Indjeian VB et al. E2F mediates cell cycle-dependent transcriptional repression in vivo by recruitment of an HDAC1/mSin3B corepressor complex. Genes Dev 2002; 16:933-947.
206. Riley DJ, Liu CY, Lee WH. Mutations of N-terminal regions render the retinoblastoma protein insufficient for functions in development and tumor suppression. Mol Cell Biol 1997; 17:7342-7352.
207. Robanus-Maandag E, Dekker M, van der Valk M et al. p107 is a suppressor of retinoblastoma development in pRb-deficient mice. Genes Dev 1998; 12:1599-1609.
208. Ruas M, Peters G. The p16INK4a/CDKN2A tumor suppressor and its relatives. Biochim Biophys Acta 1998; 1378:F115-177.
209. Saavedra HI, Wu L, de Bruin A et al. Specificity of E2F1, E2F2, and E2F3 in mediating phenotypes induced by loss of Rb. Cell Growth Differ 2002; 13:215-225.
210. Sage J, Mulligan GJ, Attardi LD et al. Targeted disruption of the three Rb-related genes leads to loss of G(1) control and immortalization. Genes Dev 2000; 14:3037-3050.
211. Schneider JW, Gu W, Zhu L et al. Reversal of terminal differentiation mediated by p107 in Rb -/- muscle cells. Science 1994; 264:1467-1471.
212. Sears R, Ohtani K, Nevins JR. Identification of positively and negatively acting elements regulating expression of the E2F2 gene in response to cell growth signals. Mol Cell Biol 1997; 17:5227-5235.
213. Sellers WR, Rodgers JW, Kaelin Jr WG. A potent transrepression domain in the retinoblastoma protein induces a cell cycle arrest when bound to E2F sites. Proc Natl Acad Sci USA 1995; 92:11544-11548.
214. Sellers WR, Novitch BG, Miyake S et al. Stable binding to E2F is not required for the retinoblastoma protein to activate transcription, promote differentiation, and suppress tumor cell growth. Genes Dev 1998; 12:95-196.
215. Shan B, Lee WH. Deregulated expression of E2F-1 induces S-phase entry and leads to apoptosis. Mol Cell Biol 1994; 14:8166-8173.
216. Shan B, Zhu X, Chen PL et al. Molecular cloning of cellular genes encoding retinoblastoma-associated proteins: Identification of a gene with properties of the transcription factor E2F. Mol Cell Biol 1992; 12:5620-5631.
217. Sherr CJ. G1 phase progression: Cycling on cue. Cell 1994; 79:551-555.
218. Sherr CJ. Cancer cell cycles. Science 1996; 274:1672-1677.

219. Sherr CJ, Roberts JM. CDK inhibitors: Positive and negative regulators of G1phase progression. Genes & Dev 1999; 13:1501-1512.
220. Shimizu M, Ichikawa E, Inoue U et al. The G1/S boundary-specific enhancer of the rat cdc2 promoter. Mol Cell Biol 1995; 15:2882-2892.
221. Shirodkar S, Ewen M, DeCaprio JA et al. The transcription factor E2F interacts with the retinoblastoma product and a p107cyclin A complex in a cell cycle-regulated manner. Cell 1992; 68:157-166.
222. Simpson MT, MacLaurin JG, Xu D et al. Caspase 3 deficiency rescues peripheral nervous system defect in retinoblastoma nullizygous mice. J Neurosci 2001; 21:7089-7098.
223. Singh P, Coe J, Hong W. A role for retinoblastoma protein in potentiating transcriptional activation by the glucocorticoid receptor. Nature 1995; 374:562-565.
224. Smith EJ, Nevins JR. The Rb-related p107 protein can suppress E2F function independently of binding to cyclin A/cdk2. Mol Cell Biol 1995; 15:338-344.
225. Smith EJ, Leone G, DeGregori J et al. The accumulation of an E2F-p130 transcriptional repressor distinguishes a G0 cell state from a G1 cell state. Mol Cell Biol 1996; 16:6965-6976.
226. Soos TJ, Kiyokawa H, Yan JS et al. Formation of p27-CDK complexes during the human mitotic cell cycle. Cell Growth Differ 1996; 7:135-146.
227. Sterner JM, Murata Y, Kim HG et al. Detection of a novel cell cycle-regulated kinase activity that associates with the amino terminus of the retinoblastoma protein in G2/M phases. J Biol Chem 1995; 270:9281-9288.
228. Stevens C, La Thangue NB. E2F and cell cycle control: A double-edged sword. Arch Biochem Biophys 2003; 412:157-169.
229. Strobeck MW, Knudsen KE, Fribourg AF et al. BRG-1 is required for RB-mediated cell cycle arrest. Proc Natl Acad Sci USA 2000; 97:7748-7753.
230. Takahashi Y, Rayman JB, Dynlacht BD. Analysis of promoter binding by the E2F and pRB families in vivo: Distinct E2F proteins mediate activation and repression. Genes Dev 2000; 14:804-816.
231. T'Ang A, Varley JM, Chakraborty S et al. Structural rearrangement of the retinoblastoma gene in human breast carcinoma. Science 1988; 242:263-266.
232. Tao Y, Kassatly RF, Cress WD et al. Subunit composition determines E2F DNA-binding site specificity. Mol Cell Biol 1997; 17:6994-7007.
233. Thalmeier K, Snovzik H, Metz R et al. Nuclear factor E2F mediates basic transcription and transactivation by E1A of the human MYC promoter. Genes Dev 1989; 3:527-536.
234. Thomas DM, Carty SA, Piscopo DM et al. The retinoblastoma protein acts as a transcriptional coactivator required for osteogenic differentiation. Mol Cell 2001; 8:303-316.
235. Toma JG, El-Bizri H, Barnabe-Heider F et al. Evidence that helix-loop-helix proteins collaborate with retinoblastoma tumor suppressor protein to regulate cortical neurogenesis. J Neurosci 2000; 20:7648-7656.
236. Trimarchi JM, Lees JA. Sibling rivalry in the E2F family. Nat Rev Mol Cell Biol 2002; 3:11-20.
237. Tsai KY, Hu Y, Macleod KF et al. Mutation of E2f-1 suppresses apoptosis and inappropriate S phase entry and extends survival of Rbdeficient mouse embryos. Mol Cell 1998; 2:293-304.
238. Udvadia AJ, Templeton DJ, Horowitz JM. Functional interactions between the retinoblastoma (Rb) protein and Sp-family members: Superactivation by Rb requires amino acids necessary for growth suppression. Proc Natl Acad Sci USA 1995; 92:3953-3957.
239. Vairo G, Livingston DM, Ginsberg D. Functional interaction between E2F-4 and p130: Evidence for distinct mechanisms underlying growth suppression by different retinoblastoma protein family members. Genes Dev 1995; 9:869-881.
240. Vanderluit JL, Ferguson KL, Nikoletopoulou V et al. p107 regulates neural precursor cells in the mammalian brain. J Cell Biol 2004; 166:853-863.
241. Varley JM, Armour J, Swallow JE et al. The retinoblastoma gene is frequently altered leading to loss of expression in primary breast tumours. Oncogene 1989; 4:725-729.
242. Verona R, Moberg K, Estes S et al. E2F activity is regulated by cell cycle-dependent changes in subcellular localization. Mol Cell Biol 1997; 17:7268-7282.
243. Vooijs M, Berns A. Developmental defects and tumor predisposition in Rb mutant mice. Oncogene 1999; 18:5293-5303.
244. Wang ZM, Yang H, Livingston DM. Endogenous E2F-1 promotes timely G0 exit of resting mouse embryo fibroblasts. Proc Natl Acad Sci USA 1998; 95:15583-15586.
245. Weinberg RA. Tumor suppressor genes. Science 1991; 254:1138-1146.
246. Weinberg RA. The retinoblastoma protein and cell cycle control. Cell 1995; 81:323-330.
247. Weintraub H. The MyoD family and myogenesis: Redundancy, networks, and thresholds. Cell 1993; 75:1241-1244.
248. Weintraub SJ, Chow KN, Luo RX et al. Mechanism of active transcriptional repression by the retinoblastoma protein. Nature 1995; 375:812-815.

249. Welch PJ, Wang JY. A C-terminal protein-binding domain in the retinoblastoma protein regulates nuclear c-Abl tyrosine kinase in the cell cycle. Cell 1993; 75:779-790.
250. Welch PJ, Wang JY. Abrogation of retinoblastoma protein function by c-Abl through tyrosine kinase-dependent and -independent mechanisms. Mol Cell Biol 1995; 15:5542-5551.
251. Wells J, Boyd KE, Fry CJ et al. Target gene specificity of E2F and pocket protein family members in living cells. Mol Cell Biol 2000; 20:5797-5807.
252. Whitaker LL, Su H, Baskaran R et al. Growth suppression by an E2F-binding-defective retinoblastoma protein (RB): Contribution from the RB C pocket. Mol Cell Biol 1998; 18:4032-4042.
253. Whyte P, Buchkovich KJ, Horowitz JM et al. Association between an oncogene and an anti-oncogene: The adenovirus E1A proteins bind to the retinoblastoma gene product. Nature 1988; 334:124-129.
254. Williams BO, Remington L, Albert DM et al. Cooperative tumorigenic effects of germline mutations in Rb and p53. Nat Genet 1994a; 7:480-484.
255. Williams BO, Schmitt EM, Remington L et al. Extensive contribution of Rb-deficient cells to adult chimeric mice with limited histopathological consequences. Embo J 1994b; 13:4251-4259.
256. Wolffe AP, Hayes JJ. Chromatin disruption and modification. Nucleic Acids Res 1999; 27:711-720.
257. Woo MS, Sanchez I, Dynlacht BD. p130 and p107 use a conserved domain to inhibit cellular cyclin-dependent kinase activity. Mol Cell Biol 1997; 17:3566-3579.
258. Wu CL, Zukerberg LR, Ngwu C et al. In vivo association of E2F and DP family proteins. Mol Cell Biol 1995; 15:2536-2546.
259. Wu C-L, Classon M, Dyson N et al. Expression of dominant negative mutant DP-1 block cell cycle progression in G1. Mol Cell Biol 1996; 16:3698-3706.
260. Wu L, Timmers C, Maiti B et al. The E2F1-3 transcription factors are essential for cellular proliferation. Nature 2001; 414:457-462.
261. Wu L, de Bruin A, Saavedra HI et al. Extra-embryonic function of Rb is essential for embryonic development and viability. Nature 2003; 421:942-947.
262. Xiao ZX, Ginsberg D, Ewen M et al. Regulation of the retinoblastoma protein-related protein p107 by G1 cyclin-associated kinases. Proc Natl Acad Sci USA 1996; 93:4633-4637.
263. Xiao ZX, Chen J, Levine AJ et al. Interaction between the retinoblastoma protein and the oncoprotein MDM2. Nature 1995; 375:694-698.
264. Xu HJ, Hu SX, Hashimoto T et al. The retinoblastoma susceptibility gene product: A characteristic pattern in normal cells and abnormal expression in malignant cells. Oncogene 1989; 4:807-812.
265. Xu HJ, Hu SX, Cagle PT et al. Absence of retinoblastoma protein expression in primary nonsmall cell lung carcinomas. Cancer Res 1991; 51:2735-2739.
266. Xu HJ, Zhou Y, Seigne J et al. Enhanced tumor suppressor gene therapy via replication-deficient adenovirus vectors expressing an N-terminal truncated retinoblastoma protein. Cancer Res 1996; 56:2245-2249.
267. Yang H, Williams BO, Hinds PW et al. Tumor suppression by a severely truncated species of retinoblastoma protein. Mol Cell Biol 2002; 22:3103-3110.
268. Yoshikawa K. Cell cycle regulators in neural stem cells and postmitotic neurons. Neurosci Res 2000; 37:1-14.
269. Zacksenhaus E, Jiang Z, Chung D et al. pRb controls proliferation, differentiation, and death of skeletal muscle cells and other lineages during embryogenesis. Genes Dev 1996; 10:3051-3064.
270. Zhang HS, Gavin M, Dahiya A et al. Exit from G1 and S phase of the cell cycle is regulated by repressor complexes containing HDAC-Rb-hSWI/SNF and Rb-hSWI/SNF. Cell 2000; 101:79-89.
271. Zhang J, Schweers B, Dyer MA. The first knockout mouse model of retinoblastoma. Cell Cycle 2004a; 3.
272. Zhang J, Gray J, Wu L et al. Rb regulates proliferation and rod photoreceptor development in the mouse retina. Nat Genet 2004b; 36:351-360.
273. Zhang Y, Chellappan SP. Cloning and characterization of human DP2, a novel dimerization partner of E2F. Oncogene 1995; 10:2085-2093.
274. Zhu L, Xie E, Chang LS. Differential roles of two tandem E2F sites in repression of the human p107 promoter by retinoblastoma and p107 proteins. Mol Cell Biol 1995a; 15:3552-3562.
275. Zhu L, Enders G, Lees JA et al. The pRBrelated protein p107 contains two growth suppression domains: Independent interactions with E2F and cyclin/cdk complexes. Embo J 1995b; 14:1904-1913.
276. Zhu L, van den Heuvel S, Helin K et al. Inhibition of cell proliferation by p107, a relative of the retinoblastoma protein. Genes Dev 1993; 7:1111-1125.
277. Ziebold U, Reza T, Caron A et al. E2F3 contributes both to the inappropriate proliferation and to the apoptosis arising in Rb mutant embryos. Genes Dev 2001; 15:386-391.
278. Zindy F, Lamas E, Chenivesse X et al. Cyclin A is required in S phase in normal epithelial cells. Biochem Biophys Res Commun 1992; 182:1144-1154.

Chapter 9

Rb and Cellular Differentiation

Lucia Latella and Pier Lorenzo Puri*

The pivotal role of the Retinoblastoma gene product p110 (pRb) in cellular differentiation has been postulated since the identification of pRb as a target of oncogenic events.[1-3] The demonstration of the essential role of pRb during terminal differentiation of many tissues appeared evident along with the identification of the critical properties of pRb in regulating cell cycle progression and apoptosis.[4-6] Permanent cell cycle arrest and resistance to both tumor formation and apoptosis are three cardinal features of terminal differentiation.[7-11] Upon functional or genetic inactivation of pRb, cells from skeletal muscle, neuronal and hematopoietic lineages exhibit higher extent of apoptosis, fail to permanently exit the cell cycle and show incomplete differentiation.[12-24]

The mechanism by which pRb regulates cellular differentiation has been addressed primarily in skeletal muscle cells. Skeletal myogenesis can be recapitulated in vitro by exploiting the ability of several muscle cell lines to differentiate into multinucleated myotubes. This experimental differentiation system offers the unique possibility to investigate the events that regulate at the molecular level the expression of differentiation genes. Using this system, a number of studies have uncovered the essential role of pRb in the activation of muscle-gene transcription, and demonstrated that this additional function of pRb is independent on its ability to regulate cell cycle and apoptosis.

While most of the biological functions of pRb are shared with two closely related mammalian proteins—p107 and pRb2/p130,[4,25,26]—the role of pRb in terminal differentiation of skeletal muscles, as well as other lineages, appears to be unique, as pRb could not be functionally replaced by p107 or pRb2/p130 in pRb-deficient mice.[17,21,24,27-30]

In the following paragraphs the role of pRb in the control of cell cycle, apoptosis and gene expression in cellular differentiation will be illustrated separately, with a particular emphasis on pRb function during skeletal myogenesis. The possible relationship between these apparently distinct functions of pRb will also be discussed. Furthermore, the function of pRb in extraembryonic lineages and the consequences this activity might have in survival and differentiation of certain lineages during development will be reviewed, with the aim of establishing the mechanistic insight into the cell autonomous and nonautonomous functions of pRb during embryogenesis.

Control of Permanent Cell Cycle Withdrawal by pRb during Terminal Differentiation

A typical feature of cells undergoing terminal differentiation is their ability to exit the cell cycle and their resistance to initiate DNA synthesis and proliferate in response to mitogens. For instance, muscle cells become refractory to mitogenic stimulation upon the acquisition of the differentiated phenotype.[7,8,11] During skeletal myogenesis, myoblasts exposed to differentiation

*Corresponding Author: Pier Lorenzo Puri—Dulbecco Telethon Institute, c/o Parco Scientifico Biomedico di Roma, San Raffaele, Via di Castel Romano, 100, 00128 Roma, Italy.
Email: plpuri@dti.telethon.it

Rb and Tumorigenesis, edited by Maurizio Fanciulli.

cues, which can be recapitulated by serum withdrawal in culture conditions, cease to divide, and differentiate into multinucleated myotubes, whose nuclei are irreversibly confined into a post-mitotic state.[8,9,31] pRb, as well as the other two "pocket proteins" (p107 and pRb2/p130), are pivotal regulators of the cell cycle machinery that governs G1-S progression and DNA synthesis.[25] During cellular proliferation, serum-activated cyclin/cyclin-dependent kinase (cyclin/ cdk) complexes phosphorylate pRb, p107 and pRb2/p130 to disrupt interactions with members of the E2F family, thereby inactivating the anti-proliferative activities of these "pocket proteins". Heterodimers formed by E2F proteins and DP family members, once released from the interaction with pRb-family members, stimulate the expression of genes essential for cell cycle progression and DNA synthesis, via binding to the regulatory elements of these genes.[4,25,26] Upon serum withdrawal, or following other differentiation cues, the simultaneous decline of cyclin/cdk complex expression and increased levels of their inhibitors—the cyclin-dependent kinases inhibitors (cdki)—causes rapid de-phosphorylation of pocket proteins.[7,32-34] Among them, only pRb is significantly up-regulated in differentiating myoblasts, through a MyoD-mediated transcriptional induction.[15,29,35] Increased nuclear affinity and peculiar distribution of pRb in the nuclei of myotubes has been also described.[36,37] Hypophosphorylated pRb has increased affinity for E2F/DP heterodimers, and the recruitment of pRb on E2F-responsive sites represses the transcription of target genes, leading to the arrest of the cell cycle at the G0/G1 phase – an event necessary for the initiation of the myogenic program.[7,9,11,38,39] Although the precise distribution and temporal sequence of "pocket protein" recruitment on E2F-responsive promoters is not completely known, a number of studies indicate that in myoblasts entering the quiescence status, prior to the expression of muscle genes, the formation of pRb2/p130-E2F4 complexes precedes the formation of pRb-E2F complexes, which are typically abundant only at late stages of terminal differentiation.[40-43] Interestingly, a sub-set of myoblasts, which remain quiescent and undifferentiated, and retain the ability to self renew—the "so called" reserve cells—show the predominant expression of pRb2/p130, as the preferential E2F-binding partner.[44]

The molecular and biochemical detail underlying the ability of pRb to repress the transcription of E2F target genes has recently been elucidated. pRb can inactivate E2F-dependent transcription by at least two distinct mechanisms. First, the direct association of pRb with the transactivation domain of E2F prevents the recruitment of the basal transcription machinery to the E2F-responsive promoters.[4,6,25] Second, the pRb-mediated recruitment of histone deacetylases to E2F binding sites, causes promoter hypoacetylation and the subsequent recruitment of the histone methyltransferase SUV39H1 and the heterochromatin protein 1 (HP1) to silence E2F-responsive promoters.[4,25,45,46] Furthermore, other proteins belonging to complexes that silence transcription by modifying the chromatin have been described to interact wi' pRb, including Brg1, the enzymatic sub-unit of the SWI/SNF chromatin remodeling c plex, and member of the Polycomb complex.[47,48] As the ability to recruit histone deace on E2F-responsive promoters was also described for pRb2/p130 and p107, it is likely myoblasts entering the state of quiescence, the formation of complexes containing E bers, 'pocket proteins" and histone deacetylases mediates the repression of tran those genes that promotes G1/S phase progression.[4]

The regulation of the cell cycle during skeletal myogenesis can be separ distinct stages. The cell cycle arrest at the G0/G1 boundary in response to di (mimicked in culture conditions by mitogen deprivation), the establishme cell cycle withdrawal, and the maintenance of the post-mitotic state in ated myotubes.[8] The ability of viral oncoproteins, such as adenovirus (TAg), papillomavirus (E7) and Polyomavirus T antigen, to interfere v by disrupting the formation of "pocket proteins" and E2F complexe elucidate the role of pRb in the regulation of the cell cycle during s viral proteins share the ability to interfere with the myogenic pro cultured in low mitogen conditions from exiting the cell cycle.[12,

n de inc the and exit o up-str stimul

also repress the expression of muscle-specific genes, although the extent of repression of differentiation varies among these proteins.[12,13,19,22,23,49-52] The anti-myogenic activities of E1A and SV40 TAg partly depend on their ability to interact and inhibit the pro-differentiation functions of p300/CBP acetyltransferases.[13,23,50] As both pocket proteins and acetyltransferases control the initial cell cycle arrest occurring in differentiating myoblasts, it is possible that sustained proliferation might be responsible for the inhibition of differentiation upon forced expression of viral oncoproteins in myoblasts. Indeed, over-expression of upstream negative regulators of the pRb pathway (e.g., cyclins), nuclear inhibitors of pRb (e.g., Id and EID) or pRb downstream targets (e.g., E2F1) is sufficient to impair the differentiation program in muscle cells.[38,53-57] However, it is still unclear whether sustained proliferation alone is sufficient to prevent the expression of differentiation genes in muscle cells, as expression of differentiation markers in the presence of DNA synthesis has been reported in myoblasts under particular experimental conditions.[53] Thus, it is likely that pRb might regulates cell cycle and muscle gene expression by distinct activities (see paragraph below).

The unique role of pRb in establishing the post-mitotic state in terminally differentiated myotubes is supported by strong experimental evidence. The pRb-binding region is required for the viral oncoprotein E1A to enforce G1-S phase progression and stimulate DNA synthesis in terminally differentiated myotubes.[23,50] The same phenotype was observed in cultured myotubes derived from pRb-deficient myoblasts, but not in p107 or pRb2/p130 null myotubes.[21] Notably, in neither myotubes ectopically expressing E1A or in myotubes derived from pRb-/- myoblasts, DNA synthesis is followed by progression through G2-M phase and cell division, due to mitotic catastrophe,[20,21,24,34,58] although aberrant mitosis of nuclei from myotubes infected with E1A has been reported.[59] In contrast, p300/CBP and the other "pocket proteins", pRb2/p130 and p107 appear to be important for the cell cycle arrest prior to the expression of muscle genes, but dispensable for the establishment of the post-mitotic phenotype.[23,24,28,50,60,61] Interestingly, whether the property of pRb to prevent aberrant reactivation of the cell cycle in myotubes relies on its ability to bind E2F proteins and repress the transcription of E2F target genes is still controversial, as some reports showed that overexpression of E2F proteins is not sufficient to reactivate DNA synthesis in normal myotubes and does not recapitulate the phenotype of pRb-deficient myotubes,[61,62] but others reported DNA synthesis in muscle cells overexpressing E2F1.[63] The interpretation of the role of pRb during in vivo differentiation is complicated by the fact that pRb null mice are not viable.[14,16,17] However, experiments performed in pRb-/- embryos rescued by transgenic expression of low levels of pRb, driven by a pRb mini-gene, which allows pRb null mice to survive to birth, revealed specific defects in skeletal muscle differentiation.[24] Among them, it was observed DNA synthesis in myotube nuclei, with an increased extent of endoreduplication and apoptosis. This phenotype can be explained by an aberrant reenter into the cell cycle, followed by apoptosis, in nuclei from myofibers with reduced pRb levels. Moreover, these defects were exacerbated by the combined inactivation of p21.[64] These results highlight the importance of pRb and p21 in regulating the irreversible withdrawal from the cell cycle in terminally differentiated muscle cells.

Consistent with their pivotal role in maintaining the post-mitotic phenotype in myotubes, the transcription of both pRb and p21 is stimulated by MyoD at the early onset of differentiation and their protein levels remain elevated during the whole differentiation program and in ultinucleated myotubes.[15,32,33,35] High levels of these proteins inactivate the cell cycle achinery and ensures the permanent exit from the cell cycle, which is accompanied by wnregulation and functional inactivation of several MyoD-inhibitors expressed in myoblasts, uding cyclins and cdks, Id proteins, EID, HDACs and Ras.[9,34,39,54,61,65,66] According to i-directional nature of the interplay between myogenic regulatory factors (such as MyoD) ell cycle regulators (such as the cdki p21 and pRb) in regulating the permanent cell cycle differentiating muscle cells, it is likely that p21 lies downstream to MyoD, and functions am to pRb.[34] Indeed, the resistance of terminally differentiated myotubes to mitogenic appears to rely on their inability to form active cyclin/cdk complexes in response to serum, resulting in the failure to sustain pRb phosphorylation.[11,32,33] This block can be explained by the combined effect of high levels of p21 (and possibly other cdks) and low levels of cyclins in myotubes exposed to mitogens – a feature that distinguishes myotubes from undifferentiated quiescent myoblasts, which instead can form active cyclin/cdk complexes and reenter the cell cycle in response to serum.[7,11,23,34,42] Furthermore, high levels of MyoD can block the cell cycle progression via direct interaction with cyclin/cdk complexes, and inhbition of their enzymatic activity.[39] This acute block of the cell cycle by MyoD can occur independent on p21; however, the role of pRb in MyoD-dependent cell cycle arrest remains unknown.

In contrast to other vertebrates, adult urodele amphibians, such as the newt, can regenerate their limbs via mitotic reactivation and local reversal of differentiation. This process is well illustrated by elegant experiments of Brockes and coworkers, showing that newt myotubes can be induced to enter S phase by serum-contained thrombin, which triggers cdk-dependent pRb phosphorylation.[67-69] The importance of cyclin/cdk repression in mantaining the post-mitotic state of myotubes is also revealed by experiments showing that forced expression of cyclin D1 and cdk4 in myotubes can duplicate the effect of E1A, that is inducing pRb phosphorylation and driving nuclei into S phase, albeit also in this case G2-M phase progression is not observed.[58]

Recent evidence indicates that the role of pRb is restricted to the establishment of the cell cycle withdrawal and mitogen resistance in myotubes, and does not extend to the maintenance of the post-mitotic state in myotubes.[60,70] In two parallel studies, acute pRb deletion in either cultured myotubes or myofibers of adult mice did not cause reactivation of DNA synthesis, despite the reactivation of E2F-dependent transcription of genes leading to G1-S phase progression.[60,70] Since pRb-interaction with histone deacetylases and lysine methyltransferases is essential for the establishment of the post-mitotic state in myotubes,[45,71] it is likely that pRb promotes epigenetic modifications at particular loci (e.g., cell cycle genes), such as histone hypoacetylation and methylation, leading to chromatin condensation and formation of heterochromatin, which eventually persist in the absence of pRb. By contrast, acute excision of pRb in cultured adult mouse myoblasts was sufficient to deregulate the cell cycle and disrupt the differentiation process.[70]

Likewise, the anti-proliferative activity of pRb provides a cardinal function in the establishment of the differentiation program in several other tissues.[6] Interestingly, the role of pRb in regulating the cell cycle during adipocyte differentiation differs from that described in most differentiation systems.[27] A peculiar feature of adipocyte differentiation resides in the fact that, upon the hormonal induction of the differentiation program, growth-arrested preadipocytes reenter the cell cycle and undergo several rounds of clonal expansion before forming terminally differentiate adipocytes. According to this particular differentiation program, the phosphorylation status of pRb fluctuates during adipogenic differentiation. During clonal expansion of preadipocytes pRb is rendered inactive via cdk-dependent rephosphorylation, while at the later stages of adipogenesis de-phosphorylation of pRb is necessary for terminal differentiation and the acquisition of the post-mitotic state.[72] The ability of pRb to promote the expression of differentiation genes is restricted to the very late stages of adipogenic differentiation.

Anti-Apoptotic Activity of pRb during Cellular Differentiation

The anti-apoptotic function of pRb might appear at first glance in conflict with its tumor-suppressor activity; in fact, it was discovered few years after the anti-proliferative properties of pRb were identified.

The increased apoptosis observed in those tissues, such as nervous system, skeletal muscle, lens and blood cells, that fail to differentiate in pRb-/- mouse embryos,[5,10,14,16,17,73] strongly implicated pRb as a critical anti-apoptotic gene during development. This role was confirmed by the analysis of pRb -/- transgenic mice expressing low levels of pRb driven by a minigene that could complete the embryonic development.[24] In the skeletal muscles of these embryos, massive apoptosis was observed in myoblasts, prior to their fusion into myotubes, suggesting that pRb regulates cell survival of undifferentiated myogenic cells, that is before terminal differentiation has been established. And combined inactivation of p21 exacerbated the apoptotic ratio in these fetuses, indicating a possible relationship between pRb and p21 in the control of the apoptotic machinery.[74] Moreover, the same study ruled out that the cell death observed in mice with reduced levels of pRb was dependant on the activity of E2F1 and p53, as combined inactivation of these genes, do not reduced the extent of apoptosis in muscles from these embryos. This observation indicates that the mechanism underlying pRb-dependent apoptosis in muscles differs from that operating in other tissues, which requires p53 and E2F1.[5,75] Finally, mutation of the Id2 gene, which encodes for the a helix-loop-helix transcription factors that antagonizes differentiation, rescues most of the defects of pRb-null embryos, but apparently does not reverse the apoptosis of pRb null muscles,[5,76] suggesting that the anti-apoptotic of pRb mainly relies on p21 function.

The essential role of pRb and p21 in protecting myocytes from apoptosis has been investigated in cultured muscle cells. Induction of differentiation in these cells is achieved by placing them in serum free medium. In this condition, while mitogen withdrawal removes the inhibition of muscle regulatory factors exerted by mitogen-activated cascades, thereby favoring the activation of muscle gene transcription, the concomitant interruption of serum-activated survival pathways exposes the cells to apoptosis.[10] Indeed, a considerable number of cells undergo apoptosis during skeletal myogenesis in vitro, possibly providing a mechanism of selection of differentiation-competent cells, by eliminating those cells that would eventually fail to differentiate. On the other hand, differentiation-defective cells should form a reserve population of cells available for further waves of myogenesis. Thus, in the context of muscle regeneration, pRb might play an important role in maintaining the pool of reserve satellite cells. One potential anti-apoptotic pathway in differentiating cells is triggered by the IGF-1 signalling, and entails the activation of the Pi3K/AKT cascade, leading to the induction of p21, which in turn prevents pRb phosphorylation.[32,33,56,77]

The mechanism by which pRb protects differentiation-committed progenitors against cell death is likely to be multifaceted. The correlation between apoptosis and ectopic S phase entry in neurons and muscles from pRb null embryos led to the speculation that cell death is the result of a "conflict" between proliferation and differentiation signals – that is continuation of DNA synthesis during terminal differentiation.[5] The same situation has been observed in the presence of elevated levels of E2F1, the downstream target of pRb.[53,78] Nonetheless, the role of E2F1 in apoptosis of muscles seems to be only marginal.[75] An alternative model proposes that pRb directly attenuates the apoptotic signaling and that pRb degradation by stress activated caspases is a critical event to trigger apoptosis.[74,79,80] A caspase-resistant pRb—the pRb MI mutant—in which the carboxy-terminal cleavage site has been mutated—cannot be degraded following TNFα exposure.[81] Overexpression of this mutant is sufficient to attenuate TNF-induced apoptosis in fibroblasts. Moreover, replacement of wild type pRb with the pRb MI allele in the mouse germline generates pRb MI mice, which show tissue-restricted resistance to TNF-induced apoptosis.[81] Suppression of caspase-activated apoptosis is clearly a primary function of pRb, independent of its anti-proliferative activity. However, whether this mechanism accounts for pRb-mediated resistance to apoptosis in skeletal muscles is unknown. It appears conceivable that caspase-dependent inactivation of pRb might contribute to cell proliferation as long as sufficient survival signals exist that antagonize the death function of caspases. In the absence of survival signals (e.g., mitogens), apoptosis might become predominant. Interestingly, in pRb-deficient cells the apoptotic machinery is activated by the same myogenic factors, which normally promote muscle-gene expression; a MyoD-mediated induction of apoptosis has been described in pRb null cells myoblasts.[82,83] The elevated levels of unphosphorylated pRb in myotubes appear, instead, to account for the resistance to apoptosis that is typical of myotubes.[77,84]

pRb-Dependent Expression of Differentiation-Specific Genes

A number of observations suggest that pRb exerts a pivotal role in the control of muscle gene expression during skeletal myogenesis. The lack of expression of a subset of differentiation genes in muscle cells, in which the activity of pRb was compromised, either by expression of viral oncoproteins or by genetic deficiency, strongly implicates pRb as an essential regulator of muscle-gene transcription.[20,22,23,85] That pRb-dependent control of muscle-gene expression is a function independent of its cell cycle suppressive activity is indicated by a large body of evidence. First, the observation that in differentiating myoblasts the abundance of the pRb2/p130-E2F4 complexes on the regulatory regions of E2F-dependent genes largely exceeds that of pRb-E2F complexes,[40-43] despite the up-regulation of pRb during the differentiation program[15,35] suggests that high levels of pRb can contribute to regulate tissue-specific transcription, regardless the cell cycle regulatory properties. Second, either disruption of pRb function or genetic deficiency causes specific defects of gene transcription, mainly restricted to a subset of late muscle genes, such as those responsible for the expression of contractile and structural proteins (e.g., muscle creatine kinase troponins and myosins). Genes implicated in the early stages of muscle differentiation, such as p21 and myogenin, are instead normally expressed in pRb defective or deficient cells.[13,19,20,22,23,85] Since the cell cycle withdrawal is an essential prerequisite for the expression of both early and late muscle genes, it appears evident that the regulation of the cell cycle and muscle-gene transcription occour through two separate functions of pRb. Indeed, the ability of cyclin D1 over-expression to inhibit muscle gene expression does not correlate with the inactivation of cell cycle function of pRb, as the ectopic expression of a nonphosphorylatable pRb mutant, which constitutively blocks cell cycle progression, does not reverse cyclin D1 inhibition of the myogenic program.[55] Finally, an analysis of the pRb regions necessary to promote the transcription of muscle genes indicate that pRb mutants impaired in E2F binding and associated with low risk of retinoblastoma retain the ability to activate muscle gene transcription and to promote myogenesis.[86]

As pRb is a strong repressor of transcription once recruited to target promoters,[4,25] the mechanism by which it activates muscle-gene transcription has puzzled many investigators and generated a number of hypotheses. The first evidence of an involvement of pRb in the activation of the myogenic program was provided by the demonstration that the pRb deficient SAOS cells are resistant to the myogenic conversion imposed by the ectopic expression of MyoD; however, reconstitution of pRb levels by cotransfection of pRb cDNA, was sufficient to enable MyoD-mediated myogenic conversion.[49] In a first biochemical analysis of this phenomenon, a physical direct association between MyoD and pRb was described to occur at the DNA level, within the MyoD-binding elements – the E-box. Although the functional relationship between pRb and MyoD was confirmed by further studies in pRb-deficient cells,[20,71,85,87] the physical association between these two proteins was not reproduced.[39,57,88]

A number of hypotheses have been formulated and experimentally tested to elucidate the function pRb in promoting muscle-gene transcription. A specific role for pRb has been reported in promoting functional interactions between MyoD and MEF2 to induce the expression of late muscle genes.[85] Selective inactivation of the LXCXE-binding pocket region of pRb by single aminoacid replacement, created a pRb mutant defective in recruiting histone deacetylases (HDACs).[71] This mutant revealed impaired in promoting MyoD-dependent expression of late muscle genes, and unveiled an additional function of pRb in promoting skeletal myogenesis, via disruption of MyoD-HDAC interactions, thereby relieving the inhbition of MyoD by histone deacetylases, which occurs in myoblasts.[87] In this regard, elevated levels of unphosphorylated pRb are essential to displace MyoD-HDACs interactions. Once released from HDAC, MyoD recruits several chromatin-modifying complexes (e.g., acetyltransferases and SWI/SNF complex) to activate muscle-gene transcription.[46] It is possible that additional signals, such as acetylation also contribute to regulate pRb ability to promote skeletal myogenesis.[89]

Finally, genetic evidence demonstrates that increased Ras activity is responsible for the lack of late muscle gene expression in pRb null muscles.[65,66] Ras is a pivotal mediator of mitogenic signals that antagonize the myogenic program by inducing pRb hyperphosphorylation, but also via pRb-independent mechanisms.[9,90] The fact that the Ras pathway is deregulated in the absence of pRb indicates that the relationship between pRb and mitogenic pathway is bi-directional and makes an essential contribution to the regulation of skeletal myogenesis. However, the relationship between Ras activation, pRb function and deacetylase-dependent interference with the myogenic program is currently unknown.

Notably, pRb-dependent activation of tissue-restricted gene transcription has been described in other differentiation system – e.g., adipogenesis and osteogenesis,[17,91-92bis] indicating the role of pRb as a general activator of gene expression in differented cells.

Relationship between the Abilities of pRb to Regulate Cell Cycle, Apoptosis and Gene Expression, and pRb Function in Extraembryonic Lineages

The crucial role of pRb in cell cycle control, differentiation and survival, and the fact that pRb is found inactivated in most human cancer reflects the importance of this gene during embryonic development.

Rb-/- embryos die between embryonic days (E) 13.5 and 15.5 showing defects in neurogenesis, erythropoiesis, muscle and lens development.[14,16,17] These lineages are able to initiate the differentiation programs, yet they fail to complete the differentiation process. As illustrated in the previous paragraphs, ectopic cell cycle entry and increased apoptosis characterize several tissues in Rb-/- embryos. Because germline Rb-/- embryos die too early to complete terminal differentiation in many tissues, a number of studies were performed by conditionally mutating Rb in specific tissues in order to study the role of Rb in development and differentiation. So far, rescue of embryonic lethality of Rb null mice was obtained by chimeric approaches.[93,94] More recently, chimeric mice made of wild type and Rb-/- cells were generated.[95] These studies revealed that Rb-/- chimeras are viable, fertile and display minor abnormalities compared to their germline Rb-/- counterpart. For instance, although brains of midgestation chimeric embryos display ectopic S-phase entry with increased p53 activity and elevated expression of the p53 target, p21, leading to a G2 arrest, the presence of wild type Rb allows cells to survive and to differentiate toward the neuronal fate.

These findings indicate the existence of an additional, non-cell intrinsic, function of Rb that is implicated in the regulation of cell differentiation, cell cycle and survival of certain tissue. In other words pRb can exert a non-cell-autonomous function during embryo development. Gene function is considered cell autonomous when a cell displays a phenotype corresponding to its genotype, regardless of the genotype of surrounding cells. The phenotype dysplayed by chimeric mice made of wild type and pRb null cells indicate that the cross talk between the Rb-/- cells and their wild type surrounding counterpart enables the former to survive under conditions that would otherwise be lethal in germline Rb-/- embryos. The absence of complete cells division in Rb-/- chimeras suggest that survival in this setting may results from cell cycle arrest in G2 prior to entering mitosis. Differently, the whole Rb knockout undergoes complete cell division leading to apoptosis.

Specific deletion of the pRb gene restricted to the telencephalon, has permitted the evaluation of the role of Rb in cortical development in the absence of other embryonic defects.[96] These mutant mice displayed ectopic cell divisions similar to whole Rb knockouts and different from Rb chimeras, in which cells arrest in G2. Despite the ectopic cell division in telencephalon, loss of Rb is not associated to massive neuronal apoptosis.[96] This enhanced cellular survival correlated with increased expression of TuJ1, an early neuronal marker. These data indicate that embryonic defects and not deregulated cell division are responsible for the neuronal cell death in Rb-/- knockout mice.

The hypoxia caused by defective erythropoiesis in Rb null mice can also contribute to tissue-specific apoptosis. A Cre-lox system, in which Rb was deleted only in the central nervous system (CNS), in the peripheral nervous system (PNS) and in the lens, was instrumental to define the function of pRb during neurogenesis in the presence of normal erythropoiesis.[18] The phenotype of these mice show that the conditional mutants are resistant to apoptosis in CNS, but not in PNS and lens, indicating that hypoxia contributes to cellular death of CNS neurons during developing Rb-/- mice.

Furthermore, the observation that Rb-/- mice show evident defects in the labyrinth layer of the placenta associated to a decreased nutrient transport through the placental from mother to fetus prompted further studies aimed at investigating the function of pRb in extraembryonic lineages. These studies tested the possibility that certain tissue-specific defects described in pRb-null mice could be attributed to a defective placenta development (non-cell autonomous function of pRb), rather than being ascribed to a cell-intrinsic defective function of pRb. Wild type placenta was supplied by tetraploid aggregation or by genetic manipulation to pRb null embryos.[93-95,97,98] In this setting, Rb-/- embryo survived until birth and no abnormalities on neuronal and erythroipoietic development neither apoptosis in the CNS could be detected. From these data, it can be extrapolated that a non-cell autonomous function of pRb contributes to neurogenesis and erytropoiesis. By contrast, rescued pups display defects in skeletal muscle development, indicating that a cell-intrinsic role of Rb is necessary for skeletal muscle differentiation.

The intrinsic function of Rb in inducing skeletal differentiation during development, is further supported by two recent reports showing the absolute requirement of Rb for progression through myogenic differentiation, using the muscle-specific deletion of pRb.[60,70]

A cell intrinsic function during erythropoiesis has been further questionned in a latest report, carried out by analyzing acute Rb deletion in vitro and under different stress conditions in vivo; these experiments shows that loss of Rb lead to a defective erythropoiesis, and indicate a cell-specific requirement of the pRb gene for differentiation of this particular lineage.[73] Thus, whether pRb has a cell-intrinsic function in erytropoiesis remains controversial.

Taken together, these data indicate a dual function for Rb: a cell autonomous role of pRb in the differentiation of certain tissues during development (i.e. skeletal differentiation); and a noncell autonomous function, probably implicating an extraembryonic function of Rb, that is essential for survival and induction of neuronal differentiation.

A specific survival factor, an anti-apoptotic signal or a better growth environment could explain the rescue of Rb noncell-autonomous functions. This signal could be a secreted factor(s) that protect Rb-/- cells from cellular death enabling them to differentiate. Another possibility is that, when wild type and mutant cells coexist, an apoptotic promoting factor become diluted below a critical threshold level.

References

1. Hamel PA, Phillips RA, Muncaster M et al. Speculations on the roles of RB1 in tissue-specific differentiation, tumor initiation, and tumor progression. FASEB J 1993; (10):846-54.
2. Wiman KG. The retinoblastoma gene: Role in cell cycle control and cell differentiation. FASEB J 1993; (10):841-5.
3. Yee AS, Shih HH, Tevosian SG. New perspectives on retinoblastoma family functions in differentiation. Front Biosci 1998; 3:D532-47, (Review).
4. Blais A, Dynlacht BD. Hitting their targets: An emerging picture of E2F and cell cycle control. Curr Opin Genet Dev 2004; 5:527-32.
5. Chau BN, Wang JY. Coordinated regulation of life and death by RB. Nat Rev Cancer 2003; (2):130-8.
6. Classon M, Harlow E. The retinoblastoma tumour suppressor in development and cancer. Nat Rev Cancer 2002; (12):910-7.
7. Lassar AB, Skapek SX, Novitch B. Regulatory mechanisms that coordinate skeletal muscle differentiation and cell cycle withdrawal. Curr Opin Cell Biol 1994; (6):788-94.
8. Nadal-Ginard B. Commitment, fusion and biochemical differentiation of a myogenic cell line in the absence of DNA synthesis. Cell 1978; (3):855-64.

9. Puri PL, Sartorelli V. Regulation of muscle regulatory factors by DNA-binding, interacting proteins, and post-transcriptional modifications. J Cell Physiol 2000; (2):155-73.
10. Walsh K. Coordinate regulation of cell cycle and apoptosis during myogenesis. Prog Cell Cycle Res 1997; 3:53-8.
11. Walsh K, Perlman H. Cell cycle exit upon myogenic differentiation. Curr Opin Genet Dev 1997; (5):597-602.
12. Webster KA, Muscat GE, Kedes L. Adenovirus E1A products suppress myogenic differentiation and inhibit transcription from muscle-specific promoters. Nature 1988; 332(6164):553-7.
13. Caruso M, Martelli F, Giordano A et al. Regulation of MyoD gene transcription and protein function by the transforming domains of the adenovirus E1A oncoprotein. Oncogene 1993; (2):267-78.
14. Clarke AR, Maandag ER, van Roon M et al. Requirement for a functional Rb-1 gene in murine development. Nature 1992; 359(6393):328-30.
15. Endo T, Goto S. Retinoblastoma gene product Rb accumulates during myogenic differentiation and is deinduced by the expression of SV40 large T antigen. J Biochem (Tokyo) 1992; 112(4):427-30.
16. Jacks T, Fazeli A, Schmitt EM et al. Effects of an Rb mutation in the mouse. Nature 1992; 359(6393):295-300.
17. Lee EY, Chang CY, Hu N et al. Mice deficient for Rb are nonviable and show defects in neurogenesis and haematopoiesis. Nature 1992; 359(6393):288-94.
18. MacPherson D, Sage J, Crowley D et al. Conditional mutation of Rb causes cell cycle defects without apoptosis in the central nervous system. Mol Cell Biol 2003; (3):1044-53.
19. Maione R, Fimia GM, Holman P et al. Retinoblastoma antioncogene is involved in the inhibition of myogenesis by polyomavirus large T antigen. Cell Growth Differ 1994; (2):231-7.
20. Novitch BG, Mulligan GJ, Jacks T et al. Skeletal muscle cells lacking the retinoblastoma protein display defects in muscle gene expression and accumulate in S and G2 phases of the cell cycle. J Cell Biol 1996; (2):441-56.
21. Schneider JW, Gu W, Zhu L et al. Reversal of terminal differentiation mediated by p107 in Rb-/- muscle cells. Science 1994; 264(5164):1467-71.
22. Tedesco D, Caruso M, Fischer-Fantuzzi L et al. The inhibition of cultured myoblast differentiation by the simian virus 40 large T antigen occurs after myogenin expression and Rb up-regulation and is not exerted by transformation-competent cytoplasmic mutants. J Virol 1995; 69(11):6947-57.
23. Tiainen M, Spitkovsky D, Jansen-Durr P et al. Expression of E1A in terminally differentiated muscle cells reactivates the cell cycle and suppresses tissue-specific genes by separable mechanisms. Mol Cell Biol 1996; (10):5302-12.
24. Zacksenhaus E, Jiang Z, Chung D et al. pRb controls proliferation, differentiation, and death of skeletal muscle cells and other lineages during embryogenesis. Genes Dev 1996; 10(23):3051-64.
25. Dyson N. The regulation of E2F by pRB-family proteins. Genes Dev 1998; 12(15):2245-62.
26. Trimarchi JM, Lees JA. Sibling rivalry in the E2F family. Nat Rev Mol Cell Biol 2002; (1):11-20.
27. Classon M, Kennedy BK, Mulloy R et al. Opposing roles of pRB and p107 in adipocyte differentiation. Proc Natl Acad Sci USA 2000; 97(20):10826-31.
28. Cobrinik D, Lee MH, Hannon G et al. Shared role of the pRB-related p130 and p107 proteins in limb development. Genes Dev 1996; 10(13):1633-44.
29. Jiang Z, Zacksenhaus E, Gallie BL et al. The retinoblastoma gene family is differentially expressed during embryogenesis. Oncogene 1997; 14(15):1789-97.
30. Lee MH, Williams BO, Mulligan G et al. Targeted disruption of p107: Functional overlap between p107 and Rb. Genes Dev 1996; 10(13):1621-32.
31. Wei Q, Paterson BM. Regulation of MyoD function in the dividing myoblast. FEBS Lett 2001; 490(3):171-8, (Review).
32. Guo K, Wang J, Andres V et al. MyoD-induced expression of p21 inhibits cyclin-dependent kinase activity upon myocyte terminal differentiation. Mol Cell Biol 1995; (7):3823-9.
33. Halevy O, Novitch BG, Spicer DB et al. Correlation of terminal cell cycle arrest of skeletal muscle with induction of p21 by MyoD. Science 1995; 267(5200):1018-21.
34. Mal A, Chattopadhyay D, Ghosh MK et al. p21 and retinoblastoma protein control the absence of DNA replication in terminally differentiated muscle cells. J Cell Biol 2000; 149(2):281-92.
35. Martelli F, Cenciarelli C, Santarelli G et al. MyoD induces retinoblastoma gene expression during myogenic differentiation. Oncogene 1994; (12):3579-90.
36. Szekely L, Jin P, Jiang WQ et al. Position-dependent nuclear accumulation of the retinoblastoma (RB) protein during in vitro myogenesis. J Cell Physiol 1993; (2):313-22.

37. Thorburn AM, Walton PA, Feramisco JR. MyoD induced cell cycle arrest is associated with increased nuclear affinity of the Rb protein. Mol Biol Cell 1993; (7):705-13.
38. Wang J, Walsh K. Inhibition of retinoblastoma protein phosphorylation by myogenesis-induced changes in the subunit composition of the cyclin-dependent kinase 4 complex. Cell Growth Differ 1996; (11):1471-8.
39. Zhang JM, Zhao X, Wei Q et al. Direct inhibition of G(1) cdk kinase activity by MyoD promotes myoblast cell cycle withdrawal and terminal differentiation. EMBO J 1999; 18(24):6983-93.
40. Corbeil HB, Whyte P, Branton PE. Characterization of transcription factor E2F complexes during muscle and neuronal differentiation. Oncogene 1995; 11(5):909-20.
41. Kiess M, Gill RM, Hamel PA. Expression and activity of the retinoblastoma protein (pRB)-family proteins, p107 and p130, during L6 myoblast differentiation. Cell Growth Differ 1995; (10):1287-98.
42. Puri PL, Balsano C, Burgio VL et al. MyoD prevents cyclinA/cdk2 containing E2F complexes formation in terminally differentiated myocytes. Oncogene 1997; 14(10):1171-84.
43. Shin EK, Shin A, Paulding C et al. Multiple change in E2F function and regulation occur upon muscle differentiation. Mol Cell Biol 1995; (4):2252-62.
44. Carnac G, Fajas L, L'honore A et al. The retinoblastoma-like protein p130 is involved in the determination of reserve cells in differentiating myoblasts. Curr Biol 2000; 10(9):543-6.
45. Ait-Si-Ali S, Guasconi V, Fritsch L et al. A Suv39h-dependent mechanism for silencing S-phase genes in differentiating but not in cycling cells. EMBO J 2004; 23(3):605-15.
46. McKinsey TA, Zhang CL, Olson EN. Signaling chromatin to make muscle. Curr Opin Cell Biol 2002; (6):763-72.
47. Dahiya A, Wong S, Gonzalo S et al. Linking the Rb and polycomb pathways. Mol Cell 2001; (3):557-69.
48. Dunaief JL, Strober BE, Guha S et al. The retinoblastoma protein and BRG1 form a complex and cooperate to induce cell cycle arrest. Cell 1994; 79(1):119-30.
49. Gu W, Schneider JW, Condorelli G et al. Interaction of myogenic factors and the retinoblastoma protein mediates muscle cell commitment and differentiation. Cell 1993; 72(3):309-24.
50. Puri PL, Avantaggiati ML, Balsano C et al. p300 is required for MyoD-dependent cell cycle arrest and muscle-specific gene transcription. EMBO J 1997; 16(2):369-83.
51. Sacco A, Siepi F, Crescenzi M. HPV E7 expression in skeletal muscle cells distinguishes initiation of the postmitotic state from its maintenance. Oncogene 2003; 22(26):4027-34.
52. Mymryk JS, Lee RW, Bayley ST. Ability of adenovirus 5 E1A proteins to suppress differentiation of BC3H1 myoblasts correlates with their binding to a 300 kDa cellular protein. Mol Biol Cell 1992; (10):1107-15.
53. Chen G, Lee EY. Phenotypic differentiation without permanent cell-cycle arrest by skeletal myocytes with deregulated E2F-1. DNA Cell Biol 1999; (4):305-14.
54. MacLellan WR, Xiao G, Abdellatif M et al. A novel Rb- and p300-binding protein inhibits transactivation by MyoD. Mol Cell Biol 2000; (23):8903-15.
55. Skapek SX, Rhee J, Kim PS et al. Cyclin-mediated inhibition of muscle gene expression via a mechanism that is independent of pRB hyperphosphorylation. Mol Cell Biol 1996; (12):7043-53.
56. Wang J, Walsh K. Resistance to apoptosis conferred by Cdk inhibitors during myocyte differentiation. Science 1996; 273(5273):359-61.
57. Zhang JM, Wei Q, Zhao X et al. Coupling of the cell cycle and myogenesis through the cyclin D1-dependent interaction of MyoD with cdk4. EMBO J 1999; 18(4):926-33.
58. Latella L, Sacco A, Pajalunga D et al. Reconstitution of cyclin D1-associated kinase activity drives terminally differentiated cells into the cell cycle. Mol Cell Biol 2001; (16):5631-43.
59. Crescenzi M, Soddu S, Tato F. Mitotic cycle reactivation in terminally differentiated cells by adenovirus infection. J Cell Physiol 1995; 162(1):26-35.
60. Camarda G, Siepi F, Pajalunga D et al. A pRb-independent mechanism preserves the postmitotic state in terminally differentiated skeletal muscle cells. J Cell Biol 2004; 167(3):417-23.
61. Puri PL, Cimino L, Fulco M et al. Regulation of E2F4 mitogenic activity during terminal differentiation by its heterodimerization partners for nuclear translocation. Cancer Res 1998; 58(7):1325-31.
62. Pajalunga D, Tognozzi D, Tiainen M et al. E2F activates late-G1 events but cannot replace E1A in inducing S phase in terminally differentiated skeletal muscle cells. Oncogene 1999; 18(36):5054-62.
63. Gill RM, Hamel PA. Subcellular compartmentalization of E2F family members is required for maintenance of the postmitotic state in terminally differentiated muscle. J Cell Biol 2000; 148(6):1187-201.

64. Jiang Z, Liang P, Leng R et al. E2F1 and p53 are dispensable, whereas p21(Waf1/Cip1) cooperates with Rb to restrict endoreduplication and apoptosis during skeletal myogenesis. Dev Biol 2000; 227(1):8-41.
65. Lee KY, Ladha MH, McMahon C et al. The retinoblastoma protein is linked to the activation of Ras. Mol Cell Biol 1999; (11):7724-32.
66. Takahashi C, Bronson RT, Socolovsky M et al. Rb and N-ras function together to control differentiation in the mouse. Mol Cell Biol 2003; (15):5256-68.
67. Tanaka EM, Gann AA, Gates PB et al. Newt myotubes reenter the cell cycle by phosphorylation of the retinoblastoma protein. J Cell Biol 1997; 136(1):155-65.
68. Tanaka EM, Drechsel DN, Brockes JP. Thrombin regulates S-phase reentry by cultured newt myotubes. Curr Biol 1999; 9(15):792-9.
69. Velloso CP, Simon A, Brockes JP. Mammalian postmitotic nuclei reenter the cell cycle after serum stimulation in newt/mouse hybrid myotubes. Curr Biol 2001; 11(11):855-8.
70. Huh MS, Parker MH, Scime A et al. Rb is required for progression through myogenic differentiation but not maintenance of terminal differentiation. J Cell Biol 2004; 166(6):865-76.
71. Chen TT, Wang JY. Establishment of irreversible growth arrest in myogenic differentiation requires the RB LXCXE-binding function. Mol Cell Biol 2000; (15):5571-80.
72. Fajas L. Adipogenesis: A cross-talk between cell proliferation and cell differentiation. Ann Med 2003; 35(2):79-85.
73. Spike BT, Dirlam A, Dibling BC et al. The Rb tumor suppressor is required for stress erythropoiesis. EMBO J 2004; 23(21):4319-29.
74. Ho AT, Li QH, Hakem R et al. Coupling of caspase-9 to Apafl in response to loss of pRb or cytotoxic drugs is cell-type-specific. EMBO J 2004; 23(2):460-72.
75. Tsai KY, Hu Y, Macleod KF et al. Mutation of E2f-1 suppresses apoptosis and inappropriate S phase entry and extends survival of Rb-deficient mouse embryos. Mol Cell 1998; (3):293-304.
76. Lasorella A, Noseda M, Beyna M et al. Id2 is a retinoblastoma protein target and mediates signalling by Myc oncoproteins. Nature 2000; (6804):592-8.
77. Wang J, Guo K, Wills KN et al. Rb functions to inhibit apoptosis during myocyte differentiation. Cancer Res 1997; 57(3):351-4.
78. Wang J, Helin K, Jin P et al. Inhibition of in vitro myogenic differentiation by cellular transcription factor E2F1. Cell Growth Differ 1995; (10):1299-306.
79. Chen M, Wang J. Initiator caspases in apoptosis signaling pathways. Apoptosis 2002; (4):313-9.
80. Tan X, Wang JY. The caspase-RB connection in cell death. Trends Cell Biol 1998; (3):116-20.
81. Chau BN, Borges HL, Chen TT et al. Signal-dependent protection from apoptosis in mice expressing caspase resistant Rb. Nat Cell Biol 2002; (10):757-65.
82. Fimia GM, Gottifredi V, Bellei B et al. The activity of differentiation factors induces apoptosis in polyomavirus large T-expressing myoblasts. Mol Biol Cell 1998; (6):1449-63.
83. Peschiaroli A, Figliola R, Coltella L et al. MyoD induces apoptosis in the absence of RB function through a p21(WAF1)-dependent relocalization of cyclin/cdk complexes to the nucleus. Oncogene 2002; 21(53):8114-27.
84. Latella L, Lukas J, Simone C et al. Differentiation induced radio-resistance of muscle cells. Mol Cell Biol 2004; (14):6350-61.
85. Novitch BG, Spicer DB, Kim PS et al. pRb is required for MEF2-dependent gene expression as well as cell-cycle arrest during skeletal muscle differentiation. Curr Biol 1999; 9(9):449-59.
86. Sellers WR, Novitch BG, Miyake S et al. Stable binding to E2F is not required for the retinoblastoma protein to activate transcription, promote differentiation, and suppress tumor cell growth. Genes Dev 1998; 12(1):95-106.
87. Puri PL, Iezzi S, Stiegler P. Class I histone deacetylases sequentially interact with MyoD and pRb during skeletal myogenesis. Mol Cell 2001; (4):885-97.
88. Li FQ, Coonrod A, Horwitz M. Selection of a dominant negative retinoblastoma protein (RB) inhibiting satellite myoblast differentiation implies an indirect interaction between MyoD and RB. Mol Cell Biol 2000; (14):5129-39.
89. Nguyen DX, Baglia LA, Huang SM et al. Acetylation regulates the differentiation-specific functions of the retinoblastoma protein. EMBO J 2004; 23(7):1609-18.
90. Perry RL, Parker MH, Rudnicki MA. Activated MEK1 binds the nuclear MyoD transcriptional complex to repress transactivation. Mol Cell 2001; (2):291-301.
91. Chen PL, Riley DJ, Chen Y et al. Retinoblastoma protein positively regulates terminal adipocyte differentiation through direct interaction with C/EBPs. Genes Dev 1996; 10(21):2794-804.
92. Hansen JB, Petersen RK, Larsen BM et al. Activation of peroxisome proliferator-activated receptor gamma bypasses the function of the retinoblastoma protein in adipocyte differentiation. J Biol Chem 1999; 274(4):2386-93.

92bis. Thomas DM, Carty SA, Piscopo DM et al. The retinoblastoma protein acts as a transcriptional coactivator required for osteogenic differentiation. Mol Cell 2001; (2):303-16.
93. Maandag EC, van der Valk M, Vlaar M et al. Developmental rescue of an embryonic-lethal mutation in the retinoblastoma gene in chimeric mice. EMBO J 1994; 13(18):4260-8.
94. Williams BO, Schmitt EM, Remington L et al. Extensive contribution of Rb-deficient cells to adult chimeric mice with limited histopathological consequences. EMBO J 1994; 13(18):4251-9.
95. Lipinski MM, Macleod KF, Williams BO et al. Cell-autonomous and noncell-autonomous functions of the Rb tumor suppressor in developing central nervous system. EMBO J 2001; 20(13):3402-13.
96. Ferguson KL, Vanderluit JL, Hebert JM et al. Telencephalon-specific Rb knockouts reveal enhanced neurogenesis, survival and abnormal cortical development. EMBO J 2002; 21(13):3337-46.
97. de Bruin A, Wu L, Saavedra HI et al. Rb function in extraembryonic lineages suppresses apoptosis in the CNS of Rb-deficient mice. Proc Natl Acad Sci USA 2003; 100(11):6546-51.
98. Wu L, de Bruin A, Saavedra HI et al. Extra-embryonic function of Rb is essential for embryonic development and viability. Nature 2003; 421(6926):942-7.

Index

F

G

H

I

J

K

L

M

N

O

P

R

S

T

V

X